Improving Reliability and Quality for Product Success

Improving Reliability and Quality for Product Success

Dongsu Ryu

CRC Press
Taylor & Francis Group
Boca Raton London New York

CRC Press is an imprint of the
Taylor & Francis Group, an **informa** business

Previously published in Korean as *Jepum Hyeokmyeong and Pumjil/Shinroiseung Gisul*, by Luxmedia.

CRC Press
Taylor & Francis Group
6000 Broken Sound Parkway NW, Suite 300
Boca Raton, FL 33487-2742

First issued in paperback 2019

© 2012 by Taylor & Francis Group, LLC
CRC Press is an imprint of Taylor & Francis Group, an Informa business

No claim to original U.S. Government works

ISBN-13: 978-1-4665-0379-3 (hbk)
ISBN-13: 978-0-367-38139-4 (pbk)

Visit the Taylor & Francis Web site at
http://www.taylorandfrancis.com

and the CRC Press Web site at
http://www.crcpress.com

Contents

Preface

Improving the financial outcome when introducing new products depends primarily on the technological capability of the manufacturer, but also requires superior marketing.[*] This book focuses on the concepts and methodology for perfecting the first of the four crucial P's of marketing: product, price, place, and promotion. It is frequently said that a good product with a lower price prevails in the market. Low pricing is very important, but is not emphasized in this book, because lowering the price requires building advanced managerial systems and advanced technology development. Rather, the focus here is on creating top-quality hardware and associated durables (excluding software).

Saying a product is "good" means its value is high compared with its price. Generally, this implies its quality is higher than that of similarly priced competitors' products, or its price is lower than that of competing products of the same quality. For any product, high quality is demonstrated by trouble-free performance, customer satisfaction, and comparative performance advantages over competitors. But quality is not easy to assess, because quality concepts are complex. Explaining these concepts is easy, but managing and handling them is more difficult.

Part of the difficulty is that creating a successful product involves risk. To be a winner, a product must perform better or cost less than those of its competitors. But the better the performance and the lower the price of a preexisting product, the more design changes are needed, resulting in changing quality, which translates to a higher probability of unanticipated problems. This discourages many experienced managers from adopting new ideas. At the same time, the corporation that reacts passively toward the problems of a valuable new product will disappear from the market. Thus, eradicating such problems becomes the first priority in creating viable new products.

Product success means the market share increases sharply and lasts until competitors release responding models. Market share will also

---

[*] B. Lee, *Analysis of the Success Factor of New Product Development: Overall View*, Seoul, Korea: Industrial Technology Foundation, 2007, pp. vi, 25.

increase with product value. If the product has flaws or defects, complaints will spread through word of mouth and the Internet, and the market share will stop rising or perhaps even plummet. A sharp drop would indicate a problem with reliability procedures; a halt in increasing market share would point to insufficient use of customer satisfaction methodology and cutting-edge technology. Unfortunately, applying these solutions is problematic: reliability technology is not properly understood because its theoretical basis has not been well articulated, customer satisfaction assessment is often not done properly, and cutting-edge technology is complex because the constant introduction of new ideas widens the differences among competitors. When the market share plunges, the quality chief is also going to plunge, and he or she will start thinking about how to strengthen the quality program. But responding appropriately to such downturns can be difficult.

Let's start with reliability. A reliable product is trouble-free in use. Product performance is satisfactory and continues for the expected period. Performance depends on product design, although the technology may differ from product to product owing to unique characteristics. Generally, customers encounter few performance problems in use because designers understand the operational principles of the product well and identify and correct any issues before product release.

Nonetheless, reliability accidents occur regularly, even in Japan, which has been thought in recent years to produce the world's highest-quality products. In 2005, the media reported serious problems in two Japanese products: a charge-coupled device made by Sony[*] and Matsushita's forced flue stove.[†] Why do such failures occur in a highly advanced, technology-driven society, especially in the products of first-class corporations? The answer is that because reliability failures negatively influence business operations, they are usually kept confidential; when this information is not shared, reliability technology develops differently from company to company. Moreover, reliability theories still are not integrated among the various relevant fields. Only a multidisciplinary approach will effectively address and solve these problems. Sustaining performance through reliability technology involves different fields, with which product designers are not always familiar. Failure mechanisms—that is, the processes that lead to failure as a result of stresses on the product's

---

[*] In 2005, a major malfunction in a charge-coupled device, due to changes in materials and production in 2002, caused over 10 million cameras to lose pictures. See N. Asakawa, "Were CCDs Screened to Prevent Failure?" in *Nikkei Electronics (Japan)*, November 21, 2005, p. 97.

[†] Several deaths resulted in 2005 due to the leakage of carbon monoxide from the rubber hose of forced flue stoves made between 1985 and 1992. See T. Nishi, "Safety Analysis of Forced-Draft Balanced Flue Stove," in *Proceedings of the 38th Reliability and Maintainability Symposium*, Union of Japanese Scientists and Engineers (JUSE), 2008, pp. 179–184.

materials over its lifetime—are highly diverse. Product designers cannot be expected to thoroughly understand the science relevant to them all.

Being competitive requires either improving performance or reducing cost, which means changing either the structure or the materials of a product. These alterations can avoid or fix the relevant failure mechanisms, but may also induce new ones. Therefore, it is important to review all the various test methods, or possibly construct new ones, to identify the failures likely to affect the new structure or materials.

In most companies, the product assurance test determines whether the product will be acceptable. But even so, the product's lifetime can be difficult to predict. Sometimes a recall may be necessary within a year on a product projected to last 5 or 10 years. Therefore, reliability estimations should consider the cumulative failure rate over a specified period, rather than simply designating the product as having passed or failed product assurance testing. Testing methods for innovative products should be formulated afresh and established quantitatively. Obtaining quantitative results requires an understanding of failure mechanisms, material degradation processes, and relevant statistics. Any effort to integrate these considerations into a single approach will be inferior to an in-depth understanding of each of the three fields.

Furthermore, new failure mechanisms are often found after accidents, not during testing before product release. Many companies do not distinguish between performance technology and reliability technology, or confuse quality control checking against specifications with reliability-related quality control aimed at eliminating failures. Good quality control (inspection) systems alone can never make their products reliable. In addition, few CEOs understand the complementary relations between anticipating failure in advance and analyzing failure after the fact. Failure analyses can establish corrective actions after thorough investigation of problems that occur in the approval process or during customer use. But applying reliability assurance methods independent of the performance approval process, identifying and minimizing the various stresses that will affect the product, and establishing and executing new quantitative test methods will lead to first-class products without reliability-related accidents. All of this requires establishing a clarified definition of reliability concepts.

This book brings reliability concepts up to date, suggesting and redefining new reliability indices. I have tried to write so that laypeople as well as product engineers can easily grasp these ideas. These novel concepts can be applied to various production areas, but are particularly addressed to CEOs, because project engineers cannot properly manage problems without the support of corporate decision makers.* CEOs best

---

* G. Tellis, *Will and Vision* (Korean edition), Seoul, Korea: SIAA, 2002, pp. 345, 363.

understand the state of their businesses; they recognize the relationship of revenue accumulated to revenue planned and scheduled expenses, and therefore can project the company's surplus or deficit. They also know the exact progress of new product development and the level of commitment of the project team. But they cannot accurately predict whether the new product will become a hit or not, or whether it will continually generate many purchase orders or result in a massive recall.

The material here was previously presented to several Korean academies, and published in 2005 in the *Journal of Microelectronics Reliability* (whose chief editor is Dr. Michael Pecht, director of the Center for Advanced Life Cycle Engineering (CALCE) at the University of Maryland). In June 2006, I spoke on this topic in a one-hour plenary speech at the 36th Reliability and Maintainability Symposium organized by the Union of Japanese Scientists and Engineers (JUSE). Their report afterwards notes that this kind of back-to-basics reliability research and execution is now being implemented in Korea, while Japan is also adopting a reasonable (not quantitative) approach to reliability technology.

Chapter 1 describes design technology and product assurance. Reliability concepts, which can be confusing, are covered in Chapter 2; novel concepts of reliability and hazards originating from current technology are explained in Chapter 3. Chapter 4 discusses various product assurance tests, the practice and limits of current test methods, and the reasons why reliability failures occur repeatedly in top-brand products. Chapter 5 considers how to plan a program to appraise the reliability of final products and the methodology of using parametric accelerated life tests for responding to the problems identified. The concepts and key factors of failure analysis, which is the flip side of identifying failure, are explained in Chapter 6. How to review quality management systems and establish manufacturing quality to avoid being ambushed by unanticipated failure mechanisms, along with responding to design changes, is discussed in Chapter 7.

Core factors for new product success—the final goal in the hardware business—are considered in Chapter 8. In order to gain high market share, a new product's value must be higher than that of its competing products. A high level of quality in a new product provides better product value for a similar price. The concepts of customer satisfaction and comparative advantage over competitors, which are necessary conditions for good quality, are defined and clarified. The basic activities necessary to assess customer satisfaction and various market research methods appropriate for product development are explained, and a method to set and visualize top priorities is introduced. Finally, some methods to secure a performance edge are presented.

In Chapter 9, I propose a quantitative equation to estimate the increase in market share due to the comparative advantages of a new product.

With this equation we can predict that the number of competitors will be reduced drastically in the long run, leaving, as the final result of the competition, two strong coexisting producers. Finally, there are some comments for improving organizational competence.

The appendix offers a possible administrative policy for CEOs regarding reliability. It is crucial for CEOs to understand key reliability and quality concepts—important information for determining the direction of the operation results when the CEO knows the core questions for his employees. Thus, the goal here is that CEOs, not just specialists in product design, will be able to understand and review the effect on market share of reliability and quality. This review is no more difficult than predicting world market trends, currency rates, materials prices, and so on. While improving various production activities in accordance with the suggested methodology is the work of specialists, reviewing their results is not difficult. Quantification and visualization are key factors for good management, and through them, great advances in reliability design can be achieved, with concomitant gains in performance and customer satisfaction. These will lead to successful new product launches and consistent competitiveness.

I would be happy to receive constructive feedback on the ideas in this book. I express my deep gratitude to Yang Jaeyeol, former CEO of Daewoo Electronics, for assigning me to improve quality and reliability in its products; my colleagues at Daewoo for reviewing the material in their fields; Dr. John Evans, Institute for Advanced Engineering (Korea), for introducing various new theories and practices; Dr. Michael Pecht, Director of CALCE at the University of Maryland, for broadening reliability technology; Dr. Joo Dougyoung, President of the Korean Institute for Robot Industry Advancement, for disseminating this theory into Korean industry for over ten years; Professor Park Dongho, Hallym University, for various helpful comments; Professor Kim Myeongsoo, Suwon University, for his careful review of the text, especially on statistics, and finally, Dr. Lesley Northup, of Florida International University, for reviewing the English and logic throughout the book.

May God protect and assist us in building a global society until, in the words of the Korean national anthem, the waters of the East Sea are dried and Mt. Baekdu is worn away.

Lady Day
2009
Dongsu Ryu in Korea

section one

Introduction to reliability

chapter one

Design technology and approval specification of new products

The competition for technological innovation is now a global phenomenon. As innovation increasingly fuels a nation's wealth, all countries, especially advanced ones, are emphasizing the development of new technology. More and more, they are not only directly supporting technologically inventive companies, but also analyzing and refining their science and technology policies to enhance progress in this area.[*] In the mid-1980s, due both to Japanese technological superiority in the field of semiconductors and trade and budget deficits in the United States, there was a lot of talk about the relative decline of American supremacy, but that has diminished and virtually disappeared. Now the United States has again strengthened its global position in establishing knowledge networks. This is the result of increased efficiency in knowledge production and applications due to policies revised in the 1990s regarding innovation in science and technology.[†] Since technology is increasingly regarded as contributing to national power, other first-world nations like China are aggressively challenging the technological hegemony of the United States and Japan.

Financially, corporations generally have been taking huge up-front profits from their newly developed technologies, rather than exploring low-cost cooperative sharing of current techniques among competing firms. But it is not easy for a corporation to develop new technology alone. Consequently, the competition between individual companies has shifted to competition among systems that directly support these companies (or their extended corporate holdings) and, further, to competition among entire corporate ecosystems.[‡] Applying the term *ecosystem* to business competition implies the difficulties inherent in surviving prior to achieving prosperity—just as living things must successfully compete

---

[*] W. Song, "Study on How to Build New National Innovation Systems," Seoul, Korea: Science and Technology Policy Institute, 2004, p. 8.

[†] Y. Bae, "Knowledge, Power and Policy of Science and Technology in the U.S.A.," Seoul, Korea: Science and Technology Policy Institute, 2006, pp. 41, 35.

[‡] H. Yoon, "Advanced Model for Win-Win Cooperation between Large and Small Businesses," Special Report to Ministry of Commerce, Industry, and Energy, the Korean Association of Small Business Studies, 2006, p. 72.

for resources before becoming an established species. The synergy of all direct and indirect organizations—including not only cooperating companies and second- and third-level vendors, but also research institutes exploring future technology, colleges training personnel locally, and local and state governments backing technology development—can elevate corporations to a winning position in the global marketplace. If it is to survive in today's Darwinian competition, a corporation cannot neglect any branch of its ecosystem.

Corporations seemingly compete primarily with their products. But the true competition relates to their capacity to produce quality products. Businesses actually are engaged in a capability-building competition comprising two key factors: organizational capability and dynamic capability (responding, respectively, to inherited traits and adaptive traits in living things). Among corporations with the dynamic capability of responding properly to changing circumstances, only those with developed organizational capabilities can survive.* So far, in the capability-building competition, the most successful competitor has been Toyota of Japan,† which has the strong organizational capability to advance a product-centered strategy pursuing high quality and productivity.‡ Its customers have come to expect better quality, especially reliability, than they find in competing products, resulting in high profitability over long periods. For over a half century, Toyota made tremendous gains.§ In 2006, its sales revenue exceeded that of GM. In 2008, Toyota achieved its long-held goal of becoming the number 1 carmaker in the world, passing General Motors (the leader since 1931) in production.

Shortly after Toyota gained that distinction, however, its global sales plunged as its reputation for safety and reliability were battered by the recall of more than 8 million cars in November 2009 and January 2010.¶ This could have been predicted, based on the company's performance over the last decade. Toyota recalls reached over 1 million cars in 2003 and

---

* K. Kim, "Comparative Study on the Evolution Vector of Business Architecture between Korea and Japan," *The Korean Small Business Review*, 27(3), 2005, p. 161.

† T. Fujimoto, *Toyota Capability-Building* (Korean ed., *Noryoku Kochiku Kyoso*; Japan), Seoul, Korea: Gasan, 2005, p. 33.

‡ That is, it is in the second quadrant of the business architecture diagram described by M. Kwak, in "The Relationship between Business and Architecture and a Firm's Performance Based on the Firm Evolution Model," doctoral dissertation, the Catholic University of Korea, 2006, p. 79. Here, the X axis describes dynamic capability (+X) and organizational capability (−X) in view of corporation survival capability, and the Y axis shows the product-centered strategy (+Y) and market-centered strategy (−Y) in view of business activity.

§ D. Seok, "The Reasonable Worries of World Best Toyota," Seoul, Korea: Dong-A Ilbo, November 27, 2007, p. A34.

¶ J. Eckel, Toyota Motor Corporation, "Business Day," *New York Times*, May 7, 2010.

2 million in 2004, with a total of over 6 million between 2003 and 2006.* Subsequently, Toyota has implemented various efforts to reinforce quality by replacing certain specialists, and reliability should gradually improve.

But why does this sort of thing happen in an advanced, first-class company? Is quality degradation due to shorter development windows, cost reductions, increased work intensity, or tighter production schedules? Something must have been amiss at Toyota, which believed it possessed all the necessary capability to maintain its product advantage. Can Toyota recover its previous high quality by rehiring the quality personnel who made it great earlier? After it recovers, will Toyota's quality remain high if the quality organization is reduced again after several years due to changes in corporate policy? What can be done to advance quality in light of needed organizational decreases or cost reductions? Doubt about both Toyota's hardware and its technology has already set in. Can it be overcome?

These questions are relevant to any company that creates a product. What methods can be employed to solve these issues or, better yet, to prevent them from occurring at all? Let's start by exploring the relationship of technology to hardware product quality. In this chapter, we will examine various kinds of design technologies, their relevant characteristics and consequences, and the concept of design verification.

1.1 Design technology and manufacturing technology

You don't have to be an engineer to understand the key technology systems involved in making a product. The hierarchy of hardware technology can be outlined, from bottom to top, in the following order: service technology, manufacturing technology, design technology, and origin technology. Service technology can be acquired from the service manuals available when purchasing products. Outlines of manufacturing technology can be found through examining service manuals and performing service activities. Detailed manufacturing technology can be ascertained by observing the activities of the original equipment manufacturer (OEM). But design and origin technologies are generally kept secret. While some of them are released to the public, the types of publications describing them, such as technical papers and newly publicized standards, are inadequate.

Combining service technology as part of manufacturing technology and considering origin technology as a prior stage to design technology, hardware technology really only comprises two types: design technology and

---

* S. Kwon, "Toyota; Eruption of Quality Issues in High Growth," *Automotive Review*, Korea Automotive Research Institute, September 2006, pp. 15–19.

Table 1.1 Design Technology and Manufacturing Technology

Description	Design technology	Manufacturing technology
Concept	Product design Developing specifications for performances and manufacturing	Production according to specifications
Type	Basic theory Technical papers Testing and analysis reports Market research data	Process sequencing technology Process configuration technology Quality system
Difficulty	Difficult to understand	Easy to follow
Field	All areas of science and engineering	Mainly mechanical and electrical engineering

manufacturing technology. The former is used to design the product and decide the specifications for its performance and manufacture; the latter creates products matching those specifications. Ultimately, the hardware product is the result of manufacturing according to design specifications.

Manufacturing technology encompasses process sequencing technology, the process configuration technology requisite to each sequenced process, and the quality system; these are mainly developed and implemented by mechanical engineers. The outcomes of this technology can be seen in the production factory.

But design technology is much more complex. Deciding on specifications requires working with high-level knowledge, because the critical information for creating a good design resides in a variety of fields and levels of knowledge. This information is subsequently reported via many instruments: basic theory studies, technical papers, testing and analysis reports, market research, and so on. Integrating these widely diversified materials has to be done using a multidisciplinary approach. Moreover, to secure a competitive advantage, companies must develop unique proprietary techniques from basic scientific research, rather than relying solely on commonly used technologies. Consequently, it is not easy to understand design technology. Table 1.1 summarizes the distinctions between design and manufacturing technologies.

1.2　The substance and results of design technology

Where does the complicated design technology come from? It originates in the minds of specialists, then is made available in technical articles, specifications, reports, and drawings; finally, it is realized in actual, properly produced items. The relevant technology can be obtained by hiring experienced experts, by studying related specifications and documents, and by measuring and analyzing manufacturing equipment with high precision. Being able to produce world-best products implies that a company has acquired excellent technology in these three modes.

Design technology can be clearly understood by reference to this model. A design results from the expression of ideas conceived by scientists and engineers, published in papers, and visible in actual products, in main activities of production and verification. This process results in two essential systems, subdivided into four categories: production papers, production equipment, verifying papers, and verifying equipment. Production papers include design materials, bills of materials, drawings, and the specifications of constituent materials, components, and units—the first results of the design process. Production equipment includes the manufacturing facilities that produce parts and assembled units, which are mainly purchased from outside suppliers. (This pertains to manufacturing technology and thus is excluded from the results of design technology.) The second set of results of the design process includes verifying papers that provide approval specifications, such as testing and inspection specifications for completed products, units, components, materials, and so on, and measurement specifications for performance fundamentals. These include testing and measuring methods, sample size, sample processes, and acceptance levels, and therefore should be based on statistics. Finally, since these approval specifications must be actualized in hardware facilities, configuration equipment technology must produce verifying equipment that measures performance, checks qualities of relevant items, and so forth—the third set of results. As shown in Table 1.2, the results of design technology are thus design materials, approval specifications, and verifying equipment.

Holding the design technology for a certain item means that someone in the company establishes design materials and approval specifications, and contrives verifying equipment. If it is impossible to measure

Table 1.2 Results of Design Technology

Order	Description	Results
1	Design materials	Bills of materials, drawings, specifications of constituent materials, components, and units
—	Production equipment	Manufacturing facilities
2	Approval specifications	Measurement specifications of performance and performance fundamentals; testing and inspection specifications of assembled products, units, and components
3	Verifying equipment	Facilities to measure and inspect performance; equipment to test and analyze quality

or verify the performance and quality of products precisely after they are made, not all the relevant technologies have been addressed. Many countries designing televisions, for example, use various instruments made by the United States, Japan, or Germany, which means the companies outside those three countries are not developing their own original technologies. If, say, there is no original apparatus for checking picture quality quantitatively as one of the performance fundamentals, the company does not have the differentiation technology to make its products unique. This will damage the company's ability to survive in worldwide competition.

The final stage of technology is measuring and testing. It is very important to confirm quantitatively the value of a product's differentiation from competing products. Establishing verifying specifications and devising equipment that measures, both precisely and quickly, performance and product lifetime requires acquiring more advanced technology. It is desirable and necessary, although the challenges are great, for a company to develop its own such equipment because available instruments that are merely similar cannot accurately assess its proprietary technology. Finally, thorough and accurate evaluation specifications and facilities are as important to the final product as original design.

1.3   Basic concepts of design approval

To reiterate, securing design technology means having the capability not only to design products but also to establish testing specifications and configure testing and measuring equipment. So let's consider the concept of design approval specifications.

Product performance should match the designer's intentions and continue consistently throughout the product lifetime expected by the customer. The intended performance is confirmed by its adherence to performance specifications, and checking its invariability during use is done through reliability specifications. As new products are introduced with different performance expectations, the specifications must be revised; if the structure and materials differ from previous products, the reliability specifications must also be altered.* Because new products for the prevailing market naturally incorporate many changes, the "pass" obtained from the specifications for previous products will no longer be acceptable; obviously, the specifications applied to the old product may be inappropriate for new ones. Since the starting point of the process is valid specifications, the specifications for new or altered products must be reviewed and revised.

---

* Chapter 4, Section 4.3, discusses this in more detail.

In the 1980s General Electric (United States) and Matsushita (Japan) switched from reciprocating to rotary compressors for household refrigerators. Although a major alteration had been made in the basic structure of the refrigerator, they did not review all the related specifications. Instead, they simply accepted the previous "good" results obtained in accordance with the specifications for reciprocating compressors and launched a new refrigerator model with a rotary compressor into the market. A few years later, due to the malfunction of rotary compressors, many refrigerators were replaced with the old type, resulting in enormous financial losses. The manufacturers had assumed that the refrigerator remained essentially the same and could be checked through existing specifications—even though a core item was different. Generally, an altered product requires at least a different fixture to attach it to a tester, but in this case no changes were made in the testing. This was a fairly standard practice.

In over 30 years of manufacturing experience, I have never heard anyone express the opinion that there should be a specifications change in this case. But let's think about that. It is simply common sense that once the product is changed, the specifications should be revised. How can we apply the same specifications to a refrigerator with a new part? We cannot. Thus, all the specifications should be reviewed when a product changes, and changed to the same extent that the hardware has changed. The more change, the more revision.

Most new products are related to current models,[*] which often have lingering intractable problems. First and foremost, whether these extant issues occur or not in the new product should be determined. Next, the potential issues due to the newly designed portion should be predicted. Finally, issues of comparative disadvantage with competitors' products should be reviewed.

These three kinds of issues should be checked one by one and divided into two groups: performance issues and reliability issues. If an issue is related to materials rupture or degradation, especially over time, it is a reliability issue; if not, it is a performance issue. The specifications relevant to any problems must be explored and revised. A summary of these points is presented in Figure 1.1.[†]

Reliability is a different field than product performance. Performance can be confirmed in the present, but reliability is concerned with future problems, usually material failure due to unanticipated or accumulated stress, which is hard to identify before the product is put into

[*] If there is no current model for comparison, similar products can be researched, with the caveat that the anticipated issues would multiply, since there would be more variant components.
[†] Chapter 4, Section 4.4, discusses this in more detail.

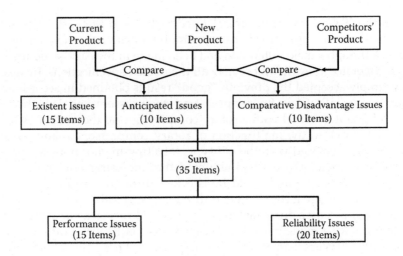

Figure 1.1 Three kinds of problems in a new product.

use. Confirming possible trouble that may occur in durables, such as automobiles or refrigerators, which are expected to need relatively little repair over 12 years, is difficult because it requires checking how much a given stress would degrade particular materials over the lifetime of the product. Thus, a series of verifying specifications should be instituted, with some of them being revised quantitatively. This implies intuitively that product designers cannot appropriately address these kinds of issues; these fall in the area of reliability, which should be managed by failure mechanism specialists with knowledge of the relevant statistics and procedures. By failing to distinguish between performance and reliability issues, many hardware producers experience serious reliability problems.

The reason for such mistakes is our confusion over the concept of reliability, which can be attributed to unsystematically defined terminology that has led to the use of different definitions from corporation to corporation and from report to report. Before mistakes occur, there are frequently misunderstandings in communication.

chapter two

The distracting jungle
of reliability concepts

Every field of knowledge, including science and technology, starts with defining its relevant terminology. With exact definitions, specialists can converse about their ideas, contributing to the development of their fields. For example, the ideas in the field of physics cannot be communicated properly if the speakers do not understand and differentiate between terms such as *force*, *energy*, and *power*. Technical terminology should be defined not only clearly, but also in terms of real phenomena, so laymen can easily approach the field.

In this regard, the terminology of reliability technology is woefully lacking. First of all, the term *reliability* itself is confusing, since it calls to mind such phrases as "good product," "no-defect product," "no-failure product," and so on. Moreover, since *high reliability* also means long-term use, the word also involves the length of the expected lifetime. It is difficult to feel confident that a product is highly reliable if the mean time to failure (MTTF), frequently used as the lifetime index, barely meets the lifetime expected by the customer. It is also difficult to differentiate reliability from the confidence level, and the varieties of relevant mathematics and seeming disconnection of reliability measures from real phenomena are dizzily confusing.

2.1   The meaning of reliability

There are several meanings of the term *reliability* in common use. First, people are considered reliable when they can receive credit from strict banks. Second, people are said to be reliable when their word can be trusted as truthful. Third, the behavior of reliable or responsible people matches what they say. Finally, since no one would have much confidence in a very old or weak person's ability to climb Mt. Everest—not because of poor planning, but because of the person's physical ability—a person is reliable (or will be successful in climbing), because his health is excellent. This last is the sense in which we use the term *reliability* here. Reliability resides not in the software, but in the hardware itself; that is, it is trouble-free in its physical substance.

Applying this to hardware products, reliability means that there are no problems with performance resulting from constituent material

Table 2.1 Korean Human Reliability Data[a]

Year	Infant mortality rate/1,000[b]	Crude death rate/1,000[c]	Life expectancy at birth (years)	Life expectancy at 60 (years)
1970	45	8	64	15
2000	9	5.2	76	20
Product-related term	Initial failure rate	Annual failure rate	BX life	

[a] Summarized from data published by the Ministry of Health and Welfare in Korea.
[b] The infant mortality rate is the yearly death rate of children under one year of age, multiplied by one thousand.
[c] The crude death rate includes the death of people at all ages. Thus, this rate differs from the concept of random failure rate in products, but is similar because there are more young and middle-aged people than there are infants and elderly.

fractures due to sudden environmental changes, such as shock, or from material degradation over time. Note that software bugs are excluded, because they are not accompanied by physical change; they pertain to performance-related issues, not to reliability. As NASA's manual of reliability training says, "Software reliability is really a misnomer" and actually means "a measure of software design adequacy."[*]

Good reliability does not mean good performance or better ease of use. For example, good reliability in an airplane means the hardware remains trouble-free during takeoff, navigation, and landing, not that the cruise speed is high or the turning radius is small.

The reliability of a product can be described in terms of two indices: lifetime and annual failure rate within lifetime, based on the occurrence of physical product disorders, or failures. To illustrate, consider human reliability indices. In this case, there are also two indices: death rate and life expectancy. The infant mortality rate, crude death rate, life expectancy at birth, and life expectancy at age 60 are common statistics reported by many countries. Table 2.1 shows the human reliability data from Korea in the years 1970 and 2000. The death rate had decreased sharply and life expectancy was much increased due to advances in health care in Korea over those 30 years. Here, the infant mortality rate and crude death rate correspond to the initial failure rate and annual failure rate of a product, respectively.[†]

Just as the health of the parents, prenatal care, and good childcare reduce the human death rate, good design, manufacture, and maintenance of hardware make products reliable. However, in the case of hardware, the task of designing and manufacturing products is more critical because the product does not have a spontaneous healing system like living things. Awkward design or improper manufacture of products can

---

[*] NASA, *Reliability Training*, 1992, p. 6.
[†] See Chapter 3, Section 3.2.

induce many initial failures at an early stage and high random failures during use, resulting in a short lifetime due to wear-out failures. The reliability target of a product should be set at zero, or nearly zero, failures for the lifetime generally expected by customers. Therefore, product reliability regards hardware as successful or reliable when the product operates without any failures during its anticipated lifetime.

Since reliability is time related, the term *future quality* can be used to describe how quality changes over time. Good reliability is good future quality. But just as every person in old age shows sign of decrepitude, every product also exhibits weaknesses over time due to degradation (wearout). Future quality should ideally be the same as initial quality, although, of course, it cannot be better than initial quality. For instance, my friend's car has been never repaired for 5 years and my car, purchased at the same time, has been repaired a few times, so my friend's car would have better future quality than mine. But that does not mean that my friend's car after 5 years of use would be better than it was when my friend purchased it. It is possible, however, to say that the future quality of a product can be better than that of competitors' products.

Reliability indices always include the concept of time. The index for lifetime includes the time dimension, but the index of the failure rate does not. The failure rate is easier to understand if duration is included, using such terms as the annual failure rate or the hourly failure rate. And the appropriate concept of time for the lifetime index also requires consideration. For example, is the mean time to failure the same as the actual lifetime?[*]

Finally, does good reliability mean a good product? A good product has high value compared to its price. This concept appears in the "survival equation" derived by the theorist S. Yoon: The product value should be greater than the product price, and the product cost should be less than the product price.[†] That is,

Product value (V) > Product price (P) > Product cost (C)

The difference between value and price is called the *consumer's net benefit*; the difference between price and cost is the *supplier's net benefit*. The supplier's net benefit becomes the absolute amount, or profit. Thus, the two elements of price are cost and profit, the proportion of which translates into the profit rate, or the index of the supplier's net benefit. The consumer's net benefit implies that the product is good, which equates to quality in the broader sense, or market-perceived quality. Therefore, the two elements of product value are price and market quality, with which the market share, or the performance index of the product value, can be estimated.[‡]

[*] See Chapter 2, Section 2.2 and Chapter 3, Section 3.4.
[†] S. Yoon, *Systemic Theory of Management (Korean)*, Seoul, Korea: GyeongMoon, 2002, p. 131.
[‡] See Chapter 9.

Since the concept of product value supersedes the concept of market quality, a good product should be interpreted as having high value. In marketing it is said that a customer buys value (the customer value or product value). This incorporates two elements: a market-perceived quality profile and a market-perceived price profile.[*] The former includes product quality and service during installation and use; the latter comprises product price, financing rate, resale price, and so on.[†]

Excluding service, good product quality can be understood by exploring three attributes that are frequently mentioned in marketing: good performance, ease of use, and trouble-free operation. In other words, a product with high market-perceived quality performs well and is well designed, easy to operate, and maintain, has no inherent defects, and is trouble-free during use. This third attribute—trouble-free use—is equivalent to good reliability, which is subsumed by the concepts of product quality and product value.

2.2 The inadequacy of the MTTF and the alternative: BX life

As we have said, the representative indices should describe lifetime and the failure rate. Indices of the latter are adequate for understanding situations that include unit periods, such the annual failure rate. But the former is frequently indexed using the mean time to failure (MTTF), which can be misleading.

The data expressed by the MTTF, which is estimated through experimentation or using applicable software, are frequently misinterpreted. For instance, assume that the MTTF of a printed circuit assembly (PCA) for a television is 40,000 h, as calculated by computer simulation. Also assume that its annual usage reaches about 2,000 h, at 5 or 6 power-on hours a day. The MTTF then equals 40,000 divided by 2,000, or 20 years, which is regarded as the lifetime of the unit. The average lifetime of the television PCA expressed with this index is thus assumed to be 20 years. But this can lead to faulty judgments and overdesign that wastes material, because actual customer experience is that the lifetime of a television PCA is around 10 years.

MTTF is often assumed to be the same as lifetime because customers understand the MTTF as, literally, the average lifetime of their appliances, so they suppose that their products will operate well until they reach the MTTF. In reality, this does not happen. By definition, the MTTF is an arithmetic mean; specifically, it equals the period from the start of

---

[*] B. Gale, *Managing Customer Value*, New York: Free Press, 1994, p. 29.
[†] Market-perceived quality is a broad concept and its constituting factors are discussed in Chapter 8.

usage to the time that the 63rd item fails among 100 sets of one production lot when arranged in the sequence of failure times.[*] Under this definition, the number of failed televisions before the MTTF is reached would be so high that customers would never accept the MTTF as a lifetime index in the current competitive market. The products made by first-class companies have far fewer failures in a reasonably defined lifetime than would occur at the MTTF. In the case of home appliances, customers expect trouble-free performance for about 10 years. After that, the failure of the TV is accepted, from the consumer's perspective. Customers would generally expect the failure of all televisions once the expected use time is exceeded—say, 12 years in the case of a television set—but they will not accept major problems within the first 10 years.

This frame of reference demonstrates that the MTTF is inappropriate as a lifetime index. So let's approach this differently. Since a zero failure rate until lifetime is nearly impossible to obtain, it is reasonable to define the lifetime as the point in time when the accumulated failure rate has reached some X%—the percent at which damage to the brand image will occur due to negative word-of-mouth reports about the product among customers. This is called the *BX life*.[†] The value of X will vary from product to product, but reaching it should be affordable for manufacturers. In the case of electrical home appliances, the time to achieve a 10 to 30% cumulative failure rate, or BX life, exceeds 10 years. In other words, the current reliability level is somewhere over 10 years, or from B10 life to B30 life. Thus, it equals an average annual failure rate of 1 to 3%, or an average annual after-sales service rate of ½ to 1½ %.[‡]

Now, from the above example, let's calculate the B10 life from the MTTF of 40,000 h. Using Equation (3.3), we get 4,000 h.[§] Since the annual usage is 2,000 h, the B10 life is 2 years, which means that the failure rate would accumulate to 10% in 2 years, and that the average annual failure rate would be 5%. The reliability level of this television, then, would not be acceptable in light of the current annual failure rate of 1 to 3%. The misinterpretation of reliability using an MTTF of 20 years would lead to higher service expenses if the product were released into the market without further improvement. A summary is presented in Table 2.2. As it shows, an MTTF of 40,000 h is equivalent to a B10 life of 4,000 h.

---

[*] The failure rate within a lifetime follows an exponential distribution. Its reliability is calculated as $R(t) = e^{-\lambda \cdot t}$, and its cumulative failure rate is $F(t) = 1 - R(t) = 1 - e^{-\lambda \cdot t} = 1 - e^{-(1/MTTF)}$, where the failure rate, λ, is $1/MTTF$. If $t = MTTF$, $F(MTTF) = 1 - e^{-1} = 0.63$.

[†] The character B in BX life comes from the German Brucheinleitzeit (initial time to fracture) or bearing (R. Abernethy, *The New Weibull Handbook*, 5th ed., North Palm Beach, FL: Abernethy, 2004, p. 2-6). In this book, the term L_{BX} represents the BX life.

[‡] See Chapter 3, Section 3.3.

[§] $L_{BX}{}^{\beta} \cong x \cdot \eta^{\beta} = (X/100) \cdot \eta^{\beta}$ (from Equation (3.3)). And in an exponential distribution, $\beta = 1$, and $\eta = MTTF$, then $L_{BX}{}^{\beta} \cong x \cdot MTTF$. Finally we get $L_{B10} \cong 0.1 \times 40,000 = 4,000$ h.

Table 2.2 Misinterpretation of Reliability Using the MTTF Index

Index	Estimation	Conversion	Appraisal	Remark
MTTF	40,000 h	20 years	Accept	Cumulative failure rate: 63%
B10 Life	4,000 h	2 years	Reject	Failure rate: 10%/2 years (Annual failure rate: 5%)

Since using the MTTF as a lifetime index breeds misunderstanding, it should not be equated with lifetime. Clearly, the BX life index more appropriately reflects the real state of product reliability.* Note that its inverse becomes the failure rate. The inverse of 20 years is an annual failure rate of 5%, which conforms with Table 2.2.

The reason for extending the lifetime to 20 years is that lifetime is calculated under the assumption that product failure is random. The annual failure rate of 1% for first-class home appliances is converted into an MTTF of 100 years, which is unbelievable. For example, televisions start to fail after 10 years, and in around 15 years nearly all sets will fail due to, for example, the degradation of the electron gun in the cathode ray tube. We know that the lifetime of a television is not 20 years, but 12 to 14 years. Since random failure cannot account for the sharply increasing failure rate after this 12- to 14-year lifetime is reached, the MTTF based on random failure or on an exponential distribution is obviously not the same as the lifetime.

2.3 Reliability and the commonsense level of confidence

Reliability has two meanings—a qualitative and a quantitative meaning. The former denotes the likelihood of item success; the latter is the probability expressing the degree of success. Reliability in the quantitative sense is defined as the ability of an item to perform a required function under stated environmental and operational conditions for a specified period of time. Thus, reliability corresponds to the probability that failure will not occur until a given time, and unreliability to the probability of failure occurring, which is equal to the cumulative failure rate; the sum of the two probabilities equals 1.[†]

---

* Perhaps the MTTF should be considered, instead, as the time for replacement or preventive maintenance. For example, assume that 100 machines are deployed and operating. At the point when 60 machines are in failure, the failed items should be replaced with new ones. As the number of failures becomes too great, the standard time for replacement should be adjusted with respect to the failure mode and its effects. Setting the target of X in the BX life is described in Chapter 3.

† Reliability ($R(t)$) and cumulative failure rate ($F(t)$) have the following relation: $R(t) + F(t) = 1$.

The confidence level denotes the adequacy of the estimated conclu-sions.* Let's say that the reliability of a certain item is 99% for 1 year, with a confidence level of 60%. The phrases "reliability of 99% for 1 year" or "cumulative failure rate of 1% for 1 year" are estimates. There is a confi-dence level of 60% that this conclusion will prove valid, or in other words, the probability of this conclusion being correct is 60%. The meaning is the same whether we use the term *assurance level* or *adequacy level*.

Let's consider the confidence level in detail. Assume that there is a population consisting of tens of thousands of products, including defec-tive ones, in which the defect rate is exactly 1%. We pick 100 items from the population at random and inspect them to find defects, then carry out this action repeatedly. Since the defect rate of the population is 1%, common sense leads us to expect to find one defect among 100 samples. But when we continue to consecutively sample groups of 100, one group after another, there are actually various findings: In one group there are no defects, in another there is one defect, in yet another two defects occur (perhaps rarely), and so on. In this case, the probability of finding one defect or more is calculated as 63%, using the Poisson distribution.† In other words, if we repeat these samplings of a hundred 100 times, we are likely to find one defect or more 63 times and to find no defects 37 times. Since we do not really repeat the sampling 100 times and actually do it only once or twice, if we find no failure, we can regard the product as good quality.

From the point of view of the consumer, the probability of correctly judging the quality of the population—that is, 63%—is called the *confi-dence level*.‡ In the above population with a 1% defect rate, the probabil-ity of finding two defects or more in a sample of 200 units is slightly

---

* In Korea and Japan, reliability has been translated into Japanese as *shinrai-sei*, and the confidence level in Japanese is *shinrai-suijun*. Using exactly the same word—*shinrai*—confuses both Japanese and Korean people. But the two concepts are totally different.

† The probability (P) of finding c or more defects is

$$P(r \geq c) = \sum_{r=c}^{\infty} \frac{(n \cdot p)^r}{r!} e^{-n \cdot p},$$

where r is the number of defects, n is the number of samples, p is the defect rate, and c is the integer nearest to $(n \cdot p)$. When $n = 100$, $p = 0.01$, and $c = 1$, then $P = 0.63$, or 63%.

‡ The lot acceptance rate (L) is given as follows:

$$L(p) = \sum_{r=0}^{c} \frac{(n \cdot p)^r}{r!} e^{-np} \leq (1 - CL),$$

where p is the LTPD (lot tolerance percent defective), c is the acceptance number, CL is the confidence level, and $(1 - CL)$ is the consumer's risk rate.

Table 2.3 Anticipated Numbers of Defects and Confidence Level

Defect rate (*p*)	Sample number (*n*)	Anticipated number of defects (*n·p*)	Probability of finding that number of defects or more	Commonsense level of confidence?
1%	100	1	63%	Yes
1%	200	2	59%	Yes
1%	300	3	58%	Yes
2%	50	1	63%	Yes
2%	100	2	59%	Yes
3%	100	3	58%	Yes
1%	200	1	86%	No
2%	200	2	91%	No
3%	200	3	94%	No

decreased to 59%, and the probability of finding three defects or more in a sample of 300 is also decreased to 58%. The term "defect rate of 1%" means 1 defect occurs among 100 sample items, 2 among 200, and 3 among 300, which is simple common sense. Therefore, I call this confidence level the "commonsense level of confidence," which can reach around 60%. Since raising the confidence level generally requires increasing the sample size, the sample size necessary to meet a confidence level of around 90%, identifying the same number of defects, would be twice the sample size using the commonsense level of confidence. A summary is in Table 2.3.

Now let's estimate defect rates conversely. Assume that there is a population for which the defect rate is not known. Also assume there are no defects when a sample of 100 is taken from the population and inspected. Is it right to conclude that its defect rate is 0%? Of course, this does not seem right—that is, the confidence level in this assumption is low. Now, again assume that there is no defect when 1,000 units from another population is sampled and inspected. Is it right to conclude then that its defect rate is 0%? The results are perplexing. The variation between the two procedures requires different conclusions. In order to create a meaningful mathematical model that accounts for this difference, let's estimate the defect rate by adding 1 to the defect number, then dividing by the number of samples. That is, if no defect is found among 100 units, the defect rate is calculated as 1% of the population, and when no defects are found in a population of 1,000, the defect rate is 0.1%. These reasonable results have a commonsense level of confidence.* If there is one defect among 100 units, the estimate of the defect rate having a commonsense level of confidence would be 2%, adding 1 to the one defect found and dividing it by 100.

Let's think about this concept from another angle. Assume that one defect is found after inspecting 100 samples of a population for which the defect rate is not known. Then the defect rate can be predicted in three cases: 1% and below, 2% and below, 3% and below. All the answers are right, but their confidence levels are different. What are the corresponding confidence levels? They would be 26, 59, and 80%, respectively. Since the estimation interval of the last case, 3% and below, is very wide, its confidence level is very high (80%), which equals the addition of 2 to the one defect. Adding 1 to the one defect, or 2% and below, has a confidence level of 59%. In the first case, without any addition, the number of defects found divided by 100—or 1%—results in a very low confidence level of 26%, which makes no sense.

Now let's review the confidence level when we add 1 to the number of defects found.† Assume that there are three populations for which defect rates are not known and that we pick up and inspect 100 samples respectively from three populations. And assume that the numbers of defects found in the three groups are, respectively, none, one, and two per 100. The confidence levels using the "add 1" estimation method—that is, 1% and below, 2% and below, and 3% and below—are then 63, 59, and 58%. These are the same as the confidence levels calculated according to the anticipated defect number in Table 2.3. Thus, the result of using the "add 1 to the number of defects found" method becomes the estimated value with a commonsense level of confidence. A summary is in Table 2.4.

---

* The upper limit of the defect rate (p) with the confidence level (CL) is estimated as follows:

$$p = \frac{\chi^2_{(1-CL)}(2r+2)}{2} \cdot \frac{1}{n}.$$

If we use in turn $CL = 0.63, r = 0$ and $CL = 0.59, r = 1$ and $CL = 0.58, r = 2$, the first chi-square term, or

$$\frac{\chi^2_{(1-CL)}(2r+2)}{2},$$

becomes ($r + 1$), or 1, 2, 3, respectively. Therefore, the defect rate with a commonsense level is given as follows: $p \cong (r + 1) \cdot 1/n$.

† In the equation for the lot acceptance rate in footnote (double cross on page 17), the term $(1 - L(p))$ becomes the commonsense level of confidence if substituting the number of failure found for c, and if p is estimated with the equation as follows: $(c + 1) = n \cdot p$. This is the statistical basis for adding one, or ($r + 1$).

Table 2.4 Confidence Level for the Estimated Value (Defect Rate)

Sample number	Defect number found (r)	(r + 1)	Estimated value	Confidence level	Commonsense level of confidence?
100	0	—	0.5% and below	39%	No
100	0	1	1% and below	63%	Yes
100	1	—	1% and below	26%	No
100	1	2	2% and below	59%	Yes
100	1	—	3% and below	80%	No
100	2	3	3% and below	58%	Yes

The commonsense level of confidence is significant because, using this concept, everyone in the organization can review and validate quality test methods with a simple calculation.

Former U.S. president Franklin Roosevelt once confessed that it was his highest aspiration that 75% of his ideas would turn out to be successful. And motivational speaker Dale Carnegie once said that being confident of 55% of his ideas would make him rich beyond any worries.* In light of this view, a confidence level of around 60%—that is, the commonsense level of confidence—is reasonably high. However, since any conclusion using the commonsense level of confidence would be right by 60% and conversely wrong by 40%, it should not be used with too much assurance. In order to find the exact confidence level of 100% and narrow down the estimation interval, validation should be performed with multilateral and in-depth scientific procedures.

Let's say that a certain company applies a confidence level of 50% to the establishment of its test methods, which means, basically, that it is half confident and half doubtful. Thus, a confidence level that exceeds 50% would be good, so it is useful to apply the commonsense level of confidence. In insurance, a confidence level of 95% would usually be applied, but in the case of product reliability, achieving such a level would cost more than twice as much as aiming for the commonsense level of confidence. This would be unacceptable. It is sufficient to start with the commonsense level of confidence to confirm new-product quality problems at an early stage. Since making mockups and managing test procedures create expenses, after identifying problems with a small sample fit to the commonsense level of confidence, it is good to increase the sample size and the confidence level. Note that if the test is repeated twice with exactly the same conditions and produces the

---

* Dale Carnegie, *How to Win Friends and Influence People* (translated into Korean), Seoul, Korea: Dale Carnegie Training, 1995, p. 180.

same results, the confidence level for that conclusion would increase to around 90%.

2.4 Dimensional differences between quality defects and failures

Generally, quality-related trouble appears for one of three reasons:

1. Customer misuse of the product
2. Nonconformance to specifications during manufacture
3. Mistakes or incompleteness in design specifications or manufacturing

An analogy of the first case would be a car crash due to the driver's dozing off. In the second instance, being out of specification or having a quality defect would obviously induce poor performance or failure in a product.* Finally, any mistakes or omissions in design or manufacture will induce faults in an otherwise high-quality system.

The word *failure* clearly indicates a physical performance disorder related to the product. The number of failures per year for a given production lot results in the annual failure rate; the time at which the same failures occur epidemically is the lifetime. The absence of failures defines good reliability.

The terms *quality defect* and *poor quality* can refer to a variety of problems that appear before customer use, such as flaws in the external case or imperfections that indicate that some malfunction will occur during use, causing poor performance or the failure to work at all. Similarly, the term *good quality* has diverse meanings. In the marketplace, the phrase connotes good aesthetic design as well as good performance. Concepts of market-perceived quality will be explored in Chapter 8, but for now, let's consider that the term *quality defect*, in the narrower sense commonly used in manufacturing, means that some aspect of the product is out of tolerance. If something can be inspected off the line and found to be out of specification, it is considered to have a quality defect. Table 2.5 shows the differences between quality defects and failures.

In the early 1990s, a company in Korea mobilized its employees to participate in a quality improvement campaign to reduce the defect rates of its components to 100 parts per million, or 0.01%; the program has now progressed to a limit of a single ppm in order to upgrade quality even further. The results of this remarkable effort are summarized in Table 2.6, which shows that the defect rate decreased by nearly one-tenth. The after-sales

---

* Sometimes no trouble occurs when the product is out of specification, which implies that the specifications are much tighter than needed. The specifications must be adjusted then, because tighter specifications generally increase manufacturing costs.

Table 2.5 Quality Defects and Failures

	Quality defect	Failure
Concept	Out of established specifications	Physical trouble
Index	Defect rate	Failure rate, lifetime
Dimension	None	1/h, year
Unit	Percent, ppm	Percent/year, year
Probability[a]	Normal distribution	Exponential distribution, Weibull distribution

[a] An exponential distribution is applied to failure rate analysis and the Weibull distribution is applied to lifetime analysis (see Chapter 3, Section 3.2). The reliabilities of the exponential distribution and of the Weibull distribution are expressed as $R(t) = e^{-\lambda \cdot t}$, $R(t) = e^{-(t/\eta)^\beta}$. In the Weibull distribution, if $\beta = 1$ and $\eta = MTTF = 1/\lambda$, then $R(t) = e^{-(t/\eta)} = e^{-\lambda \cdot t}$. In this case, the Weibull distribution becomes an exponential distribution.

Table 2.6 Results of Quality Upgrade Efforts in Company G

	Company G		Advanced company	Unit
	Year (Y)	Year (Y + 1)		
Quality defects	2,000	230	100 ~ 2,000	ppm
After-sales service rate	5,400	3,880	50 ~ 100	ppm/time
Difference	3,400	3,650	—	—

service rate was also lowered a little, contrary to the company's expectations. Many readers have noticed that the differences between quality defects and after-sales service rates never change,* and that advanced companies have surprisingly low after-sales service rates in spite of high defect rates. Why do the products of advanced companies need repair so seldom and have virtually no failures? The reason is that the concepts of defect rate and failure are completely different. Table 2.6 shows that efforts to reduce defect rates can never diminish failures. Activities to decrease quality defects for the purpose of improving reliability are as useless as trying to compare apples and oranges.

Since failure occurs even when all specifications are satisfied and no defects are evident, obviously something can be improved, by either revising specifications or making them more thorough. Because failures are principally removed by design changes or specification elaboration,

---

* Since the magnitude of the defect rate is qualitatively different from that of the after-sales service rate, the difference between the two should not be calculated. Likewise, it is inappropriate to assume any direct relationship between an initial quality study (IQS) and a vehicle dependability study (VDS), which are automobile quality indices provided by JD Power and Associates, because IQS issues pertain to quality defects, and infant failure and VDS measures wear-out failure during three years use.

reliability pertains to design technology and should be called *reliability design technology*.

Why should designs or specifications be changed? We consider failure mechanics and its two elements—stress and materials—in the next section.

2.5 The two elements of failure: Stress and material

As mentioned earlier, product failure is a physical problem. Let's focus on where the trouble occurs, or the failure site. Because every product is an aggregate of many structures, the unit structure of the failure site may only be visible when the product is disassembled. The loads applied to this unit structure produce stresses. Failure occurs when the stress is greater than the material strength, or when the material cannot endure the applied stress. Understanding this process is called *failure mechanics*,[*] and its two elements are stress and materials. A desirable or reliable structure includes both well-dispersed stresses and reliable materials. Figure 2.1 illustrates failure mechanics and its two elements.

This description sounds like it applies only to mechanical structures, but it is equally true of electronic devices, which have analogous structures. For example, semiconductors experience thin-gate rupture due to high voltages (electrical overstress (EOS)),[†] and circuit wires have

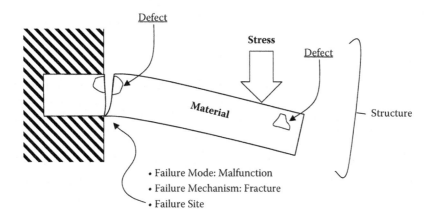

Figure 2.1 Failure mechanics.

[*] J. Evans, *Product Integrity and Reliability in Design*, New York: Springer Publishing, 2001, p. 7.
[†] EOS is a failure mechanism in which the oxide gate ruptures in response to voltage over 1,000 V/μm. M. Minges, *Electronic Materials Handbook*, Vol. 1, *Packaging*, New York: CRC Press, 1989, p. 965.

narrow-width openings due to high currents (electromigration).[*] In either case, failure occurs when the materials comprising the structures do not survive the environmental and operational stresses applied to them; after all, every object from a microdevice to a high building has some architecture, and it is only as strong as its component parts.

Because reliability is the relationship between stresses and materials, the solution for avoiding failure is altering the structures to better disperse stresses or replacing materials. But strong materials are not always reliable. Bronze is weaker than steel in mechanical strength, but better than steel in corrosive environments. It is best to select a material with consistently high mechanical strength throughout the product's usage lifetime. In Figure 2.1, the defect in the middle of the material will contribute to failure. Such defects are usually produced in the manufacturing process, which leads us to regard defects solely as a manufacturing issue. But the other defect, near the end in Figure 2.1, will not lead to failure, which implies that the specifications for flaws should be revised and expanded to include such data as location and size.

Altering the structure or materials or making specifications more detailed assumes that achieving reliability pertains to design, not manufacturing. Since product design includes reliability, it is properly done jointly by the product designer and a reliability specialist. Consequently, design specifications should be classified into two categories: performance specifications and reliability-related specifications. When an item is inspected for a certain specification and turns out to be out of tolerance, the item will not work well if the specification relates to performance; if it is a reliability specification, the item cannot reach its design target life. However, there is no need to describe specifications in detail in both the performance and reliability categories; it is often awkward or difficult to discuss each category separately in written materials about the manufacture of the item, and in many cases, it is impossible to differentiate between the two. That means that generally, the specifications of materials and structures are performance specifications as well as reliability specifications.

For example, the structures in the fuselage of an airplane should be assembled without welding because the fuselage receives repetitive stresses from the expansion and contraction of the surrounding air at high altitude and ground, which can lead to breakage along welding lines. In this case, it would be hard to distinguish whether reliability engineering was included in the design or not. Moreover, it is easy to overlook minor details that might be regarded as appropriate reliability specifications.

---

[*] Electromigration is a failure mechanism that occurs when the circuit is open if the current density is greater than 1 mA/ μm^2. Agency of Technology and Standards, *Briefs of Reliability Terminology*, Gwacheon, Korea: Ministry of Commerce, Industry, and Energy, 2006, p. 295.

For instance, the corners of plastic components are mostly produced with right angles, but giving them small round radii would make them stronger, helping to prevent failure in the short run. Likewise, the lead wires of resistors, which are used plentifully in printed circuit boards, are made of copper or brass and clad with tin for good soldering. There is a nickel layer between the two, which hampers both corrosion or the development of a galvanic cell for copper and the diffusion of zinc oxidation for brass—both of which could lead to failure due to an open circuit in a few years. Designers frequently miss minor details that are critical to reliability, like the corner radius of components and the nickel layer for soldering materials, because they seem irrelevant to performance.

Generally, failure does not occur when a product first comes out, unless the stresses are sudden and catastrophic, but only after months or years, due to gradual intrinsic changes in its materials. Thus, engineers cannot always tell ahead of production whether or not the reliability specifications are adequate. This is why product designers are often inattentive to reliability specifications. At the same time, corrective actions that require analysis of failures occurring in the market are usually considered a matter of inadequate design or manufacturing specifications. Therefore, thorough product design should include conforming product materials and their associated stress dispersal structures, and carefully assessing the operational environmental and operational stresses. This also means that the lifetime and failure rate must be predetermined, although engineers cannot estimate them before testing completed products. If a product is designed carelessly, the product lifetime will be short and its failure rate will be high, even if there is a well-organized quality control system. Thus, reliability is one of the intrinsic characteristics of a designed product. Top companies follow this practice, which is why customers prefer name brands for durables.

It is impossible to produce reliability indices of materials without considering the stresses applied to them. There is no such thing as material reliability. If the materials are incorporated into the structure and the applicable environmental/operational conditions are applied, the effects on the materials are evident and the resulting failure phenomena are observable. Generally, the reliability of a material can be determined if stresses are applied to a test piece in accordance with certain conditions, but the result actually reflects the reliability of the total structure—that is, the test piece material and the stress-producing engine. Therefore, the result cannot be applied accurately to a structure that differs from that of the test piece, with its relevant stresses. Rather, the result expresses the *relative* reliability characteristics of materials under the anticipated stresses.

As the basic concept of failure mechanics and its two elements indicates, the shape of the failure rate curve over time does not vary for different kinds of components. Electronic components consist of a combination of materials experiencing stresses, just like mechanical components. The unit structure of electronic components, including the materials and their attendant stresses, is not essentially different from that of mechanical parts, but the shape of the failure rate curve is usually assumed to be different. It is frequently said that the failure of electronic items occurs accidentally, or that failure is random, but that, for mechanical items, there are few failures within the expected lifetime and many wear-out failures near the end. Is it true that only wear-out failures occur in mechanical items and random failures in electronic products? This hypothesis was formulated in the 1950s in an attempt to explain the differences between the failure phenomena of mechanical and electronic components when the latter were just coming on the market, but it is a careless conclusion to draw today; it would be impossible to make a generalization such as this that would apply to all items. Again, let's look at an example. In the capacitors frequently used in electric power circuits, rubber caps around the terminals block the leakage of electrolytes. As the rubber degrades over time and loses elasticity, the electrolyte flows out and creates wear-out failure. Rubber degradation can be accelerated by the stress of ambient high temperature (the solution is to locate electrolyte capacitors as far as possible from heat-producing devices). There is nothing random about this phenomenon.[*] In the early industrial age, many random failures occurred in mechanical items. Nowadays mechanical items, like car engines, have little or no trouble during use and seldom fail over their lifetimes, due to over 200 years of development in mechanical engineering. If car engines are well maintained, they seldom fail.

Likewise, in the early stages of the development of electronic devices, there were many initial and random failures. Now electronic devices have been improved to the point where initial failures have mostly disappeared and random failures have decreased enormously, as the limitations applicable to circuits were observed and corrected. Pecht's law has established that the failure rates of microcircuit devices of digital systems have decreased by about 50% every 15 months.[†] It was possible to hypothesize and experimentally verify this theory because the understanding of both microstructures and the materials of electronic constituents has broadened substantially after several decades of research and failure analysis.

In brief, because failure occurs due to the correlation between stresses dispersed by the structure and the strengths of the constituent materials of the structure, a relatively high number of random failures suggests that

---

[*] See Chapter 3, Section 3.2, for a discussion of random and wear-out failure.

[†] Y. Zhang, et al., "Trends in Component Reliability and Testing," *Semiconductor International*, 1999, pp. 101–106.

engineers do not understand material behaviors and applied stresses as much as the failure rate. In nanoscale electronic devices, if the knowledge of stresses and materials was widened and deepened, there would be no random failure during use—only wear-out failure would appear over the item's lifetime, as with large buildings or automobiles. When there are more than 200 parts in an item, engineers also usually expect an exponential distribution of failure, presuming the behavior to be random. But fully understanding the structure of every item, including the connecting parts, and instituting corrective changes would ensure that only wear-out failures would occur. Consequently, failure would not follow an exponential distribution.*

Finally, the fact that hardware items are a combination of unit structures means that reliability specialists who thoroughly understand the failure mechanics of diverse unit structures can find correctives for failure, regardless of the kind of problem that produces it. Thus, reliability technology that eliminates failure is a universally applicable approach to all hardware products.

2.6 Reliability engineering as a science

So far, some confusing concepts have been clarified. These concepts include that the mean time to failure (MTTF) is inadequate as a reliability index, that failure occurs because of the relationship between materials and stress, and that corrective action for potential failures must be taken at the design stage. Given these conclusions, reliability technology as it is now practiced is virtually groundless. The Japanese call this *adopted engineering technology*,[†] meaning that reliability engineering has not been developed independently and borrows and adopts theories from other fields of science and technology. In order to establish an independent technological field, the many related topics must be addressed and integrated from the ground up. Let us briefly review the historical development of reliability engineering and the reasons why discrete reliability theories have never been established.

Research into reliability began as a result of the exceptionally high number of failures in vacuum tubes used in World War II military weapons. Over half—60 to 75%—of vacuum tubes in communications modules had failed.[‡] In the Korean War 5 years later, a similar phenomenon

---

* See Chapter 3, Section 3.2.

† Karimono Engineering, at http://www.juse.or.jp/reliability/symposium_36r&ms_repo.html.

‡ D. Kececioglu, *Reliability Engineering Handbook*, Vol. 1, Englewood Cliffs, NJ: Prentice-Hall, 1991, p. 43.

occurred. The Pentagon initiated an investigation into this enormously wasteful situation. The resulting project was undertaken by the Advisory Group on Reliability of Electronic Equipment (AGREE) and lasted for 5 years, from 1952 to 1957, when its findings were published. Subsequently, however, many of the research projects on reliability issues, such as a survey on the substitution periods of aircraft components, have been classified as confidential military documents and have not been made public.

In related fields, scientists and engineers continued to make substantial contributions on reliability issues. In the early 1950s, the basic mathematics of failure distributions was established statistically. In the 1960s, the methods called *failure modes and effects analysis (FMEA)* and *fault tree analysis (FTA)* were devised—methods that are now applied to many hardware systems. In the 1970s, there were great advances in failure analysis, mainly in materials engineering—a success that was much indebted to rapid progress in destructive testing/nondestructive testing (DT/NDT) equipment, such as the scanning electron microscope (SEM) and the scanning acoustic microscope (SAM).

Despite these achievements, since the 1970s it has been acknowledged that there is a major gap between reliability theory and its application to industrial fields—the real problem to be solved. Since the results estimated by current reliability prediction methods for electronic components differ widely from experimental data, specialists doubt the methods and their bases.* Recently, scientists have been working to clarify in detail the processes leading to failure (failure mechanisms), using the physics of failure (PoF) approach.

Why have reliability concepts and methodologies still not been established in a modern society, which has been producing so many technical advances daily? The first reason is that reliability information has not been made readily available by industries to reliability specialists. Corporations fear that releasing survey results about reliability problems will negatively influence business, so they keep them confidential. The second reason is that the solutions to reliability failures, the methodologies to determine solutions, and the foundational reliability concepts should be established through a multidisciplinary approach. The current system relegates reliability issues to the lowest level of the engineering staff or to specialists in only one field; this hampers contact among experts and the necessary interconnections among researchers in various fields. Thus, the concepts of reliability technology are not systemized and each specialist understands them idiosyncratically.

What fields comprise reliability technology? As we noted, failure occurs from the relationship between stress and material, which must

* D. Ryu, "Novel Concepts for Reliability Technology," *Journal of Microelectronics Reliability,* 45(3–4), 2006, p. 611.

be understood first and foremost. We need to know how the stresses from both the environmental conditions in which the item is operated (indoors, in a field, under the ocean, in outer space, and so on) and operational conditions (mechanical, electrical, and electronic) are concentrated or dispersed in the item's structure. We also must know how the materials in the structure reach the point of breakage, degradation, or wear due to these stresses. In order to quantify the degradation process, a model of failure mechanics must be introduced and analyzed, together with a time variable, which pertains to the field of physical chemistry. Furthermore, the manufacturing process has to be understood from the original materials to the final product, because every step influences the inner structure of the materials. Finally, knowledge of statistics is crucial for design of experiments and estimating their results. It is quite fortunate that observed results from the various interactions between stresses and materials have been investigated and publicized in the name of relevant failure mechanisms.*

After establishing the basics in the next chapter, such as that reliability comes under the larger heading of product quality, the relevant terminology will be defined, with an explication of the hierarchy of those terms. Then the systematic sequence of reliability engineering will be explored. The concepts of reliability technology are discussed and a degradation model, for which the time variable function is easy to memorize, will be connected with failure mechanisms, and the statistics for estimating lifetime will be integrated.

---

* See Chapter 6, Section 6.2.

Reliability theory and application

chapter three

Novel concepts of reliability technology

Peter Drucker has said that innovation—that is, activities increasing the value and satisfaction that consumers experience—raises the productivity and profitability of economic resources.[*] He added that to pursue higher goals, actions taken toward those goals must not simply be performed better, but must also be performed differently—that is, specific new processes that go beyond current habitual behaviors should be established. Such processes are generated by new methodologies that reflect novel concepts. A willingness to evaluate concepts that are commonly accepted uncritically can lead to new ways of considering reliability. To begin this process, various concepts need to be examined, such as the definition of reliability, the form of failure phenomena, the reliability targets for satisfactory unit items in the final product, and the statistical definition of reliability indices.

Novel concepts and the ensuing methodology should be presented clearly and simply, so CEOs can adopt these methods for product development and recognize current errors in the process. A CEO should be able to predict, before a product is released, whether it will be a success in the market or be frequently returned by customers due to quality failures. Usually, the CEO has little or no knowledge about the architecture of test specifications and test methods because the relevant information is scattered and none of it is easy to understand at a commonsense level. Neither is it easy to uncover in current texts the theory of lifetime verification described in Chapter 5. The basics of science and technology should be explicable with the application of common sense. If it is impossible to describe a given technology in ordinary language, then we really do not understand that technology well, as Richard Feynmann, the 1965 Nobel Prize winner in physics, has said.[†]

---

[*] P. Drucker, *Peter F. Drucker on Innovation* (selected articles in Korean), Seoul, Korea: Hankyung, pp. 19, 103.
[†] R. Feynmann, *Six Easy Pieces* (Korean version), Seoul, Korea: SeungSan, 2003, p. 26.

3.1 The definition of reliability and the meaning of quality

As mentioned earlier, reliability is defined as the ability of an item to perform a required function under specific environmental and operational conditions for a specified period of time. Here, the word *item* indicates a component, unit, or product. In this definition, there are two measurable variables: a specified period of time and the ability to perform. The specified period of time concerns item life, or durability. The ability to perform means the probability that the item will perform the required function, or the reliability probability. The unreliability probability, which equals 1 minus the reliable probability, becomes the cumulative failure rate. The average annual failure rate can be found by dividing the cumulative failure rate by the years in the lifetime. This confirms that determining item reliability requires clarifying two indices—item life and annual failure rate during item life. However, reports on reliability prediction frequently refer only to the failure rate, without considering the lifetime. To differentiate between types of reliability, the term *rate reliability*, abbreviated r-reliability, will be used to refer to reliability regarding only the failure rate.

So far, we have clarified that quality defects and failures are, conceptually, different dimensions of each other.[*] Thus, when we discuss good quality without quality defects and good quality with neither quality defects nor failures, the word *quality* has different meanings. To differentiate this, let's call the former conformance quality, abbreviated c-quality. Since the latter includes the concepts of both reliability and conformance to specifications, it encompasses the three different concepts of c-quality, durability, and r-reliability. The recommended unit of c-quality, which indicates inconsistency with manufacturing specifications, is a percent without dimension. The unit of durability is expressed in years, and the unit of r-reliability is the inverse of time, or percent per year. Different dimensions require different reliability activities. In brief, the design specifications for reliability should be set with the goal of extending item life until the year expected by customers, with few or no failures during the item lifetime; quality control management should aim to reduce quality defects. Table 3.1 presents this in brief.

These concepts are not clarified in the ISO/TC 16949 specifications, which all three major car makers in the United States apply to their products. Provision 4.4.4.2 under the subject "Reliability Objectives" says that "product life, reliability, durability and maintainability objectives shall be included in design inputs." Here, product life and durability both refer to item life, and reliability means item life *and* failure rate. Such terminology duplications are highly ambiguous. This may seem minor, but it is a very

---

[*] See Chapter 2, Section 2.4.

Table 3.1 The Meaning of Quality

Meaning of quality	Quality		
		Reliability	
	c-Quality	r-Reliability	Durability
Concept	Conformance to specification	Failure rate	Item life
Units	Percent	Percent/year	Year
Probability density function	Normal distribution function	Exponential distribution function	Weibull distribution function

important issue. Defining terminology correctly will help establish the reliability target clearly and quantitatively—the necessary first step for all reliability activities.

A new term, *dependability*, has been appearing recently. This is a general concept used in international standard IEC 60050-191 to explain availability and factors influencing it.* The qualitative term *dependability* is used as a general description encompassing reliability, maintainability, and maintenance support. This is also confusing. Such terminology, which is difficult to quantify, is not really helpful or useful.

If quality indices for items were expressed using indices of reliability, not unreliability, it would be easier to create the illusion that items are actually reliable. We do not really know how good item reliability is when it is expressed in terms such as 99, 99.9, or 99.99%, because in fact current products are quite reliable. Since the sum of reliability and unreliability, or the cumulative failure rate, is 1, the percentages above can be changed into cumulative failure rates of 1, 0.1, and 0.01%. An expression like this reveals significant differences in reliability in magnitudes of 10 or 100, so it is better to use the failure rate than the reliability figure. The phrase "cumulative failure rate of X% for Y years" most clearly represents the reliability of an item.

3.2   Two failure rate shapes and infant mortality distributions

Major stress due to shock ruptures materials, and cumulative minor stresses weaken materials over time, producing degradation that results in failure. The former condition is called *overstress failure* and the latter wear-out failure. Wear-out failure encompasses material deterioration

---

* Agency for Technology and Standards, *Briefs of Reliability Terminology*, Gwacheon, Korea: Ministry of Commerce, Industry, and Energy, 2006, p. 3.

due to friction, changes in the material structure due to thermal heat, and the gradual formation of new materials, such as corrosion. Because overstress failure occurs only sporadically, it seems to be random. Wear-out failure occurs incrementally over a certain period.

As we have mentioned,* the shapes of the failure rate curves for electronic devices and mechanical items do not differ. Electronic items do not exhibit only random failure, and likewise, mechanical products do not experience only wear-out failure. Items will fail because of either severe shock or wear-out caused by repetitive minor stress over time. All items, whether electronic, mechanical, or electrical, experience these two kinds of failure. As an analogy, consider the human body. The causes of death in old people and in youngsters vary widely. Old people die mainly of strokes or heart disease due to atheroma in artery walls; youngsters are more likely to die of dangerous behaviors such as speeding or high-risk activities, though both groups can be affected by any of these possibilities.

It is significant that in any given item, two kinds of failure mechanisms are operative, corresponding to failure rate and lifetime. Measures to combat overstress failures reduce random failures within a product's lifetime and corrective activities against wear-out failures extend the item life. The existence of failure mechanisms corresponding to both forms of failure clearly indicates the proper direction of activities for differentiating between the lifetime and the failure rate within the lifetime.†

Item life is determined by the weakest site, where the structural material degrades first. Since lifetime is the point when failures occur intensively at this site, wear-out failures are the principal failure phenomena. Failures aggregated into the failure rate within lifetime occur when excessive stresses from environmental and operational conditions are applied to the structural material of the item or to materials damaged in manufacturing, then incorporated into the item through faulty or inadequate screening to meet rated stresses. In this case, overstress failure is the principal phenomenon, which means that the failure rate within lifetime appears intermittently with an almost constant interval, or that failure is random.

Infant mortality appears in very unreliable items. This type of failure occurs when large stresses frequently encounter materials damaged during manufacture. In such a case, there will be many failures from the start. Infant mortality indicates failure at initial use, but also includes a type of failure rate in which many failures occur at the start of item use and then decrease rapidly over time. This is because at the beginning of item use, the damaged materials experience high stresses and consequently fail, but since the worst items fail quickly and the remaining items are not

---

* See Chapter 2, Section 2.5.
† In reliability statistics, the inverse of the failure rate indicates lifetime, but this is meaningless and inapplicable because of the different failure mechanisms described.

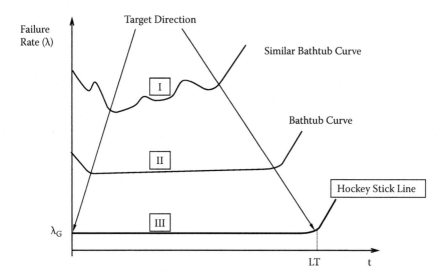

Figure 3.1 Bathtub curve and hockey stick line.

as severely damaged, the failure rate decreases. Infant mortality generally means that the product was designed without the necessary advance
investigation of environmental and operational conditions and that the
quality control system for producing the item was also unsatisfactory.

Now, if the three types of failure—infant mortality, random failure, and
wear-out failure—are connected, a bathtub curve results, as shown in the
middle curve of Figure 3.1. The failure rate for infant mortality decreases
with time, that of random failure remains constant, and the rate for wear-
out failure increases near the item's lifetime. All of these are expressed
with the Weibull distribution function.* In a Weibull distribution, the shape
parameter (β) must be understood; if it is smaller than, equal to, or bigger
than 1, then the Weibull distribution function represents decreasing infant
mortality, constant random failure (also called the *exponential function*), and
increasing wear-out failure, respectively, as shown in Table 3.2.

Since infant mortality in an item reveals major problems with item
design and manufacture, let's not pursue this issue.† A first-class product

---

* The Weibull distribution function is $R(t) = e^{-(t/\eta)^\beta}$, where η (eta) and β (beta) are the two
  decisive parameters. These are called the *scale parameter* and the *shape parameter*, respectively. The scale parameter is also called *characteristic life* and approximated to the mean
  time to failure, and the shape parameter indicates the intensity of wear-out failure.
† An item poorly addressed for reliability shows a complicated form of failure rate until
  the item reaches its lifetime. Its infant mortality includes, in part, wear-out failure; past
  the infant mortality period, various kinds of wear-out failure occur within the lifetime
  and the failure rate increases gradually over time. The item's lifetime is quite short, as
  shown in the upper curve in Figure 3.1.

Table 3.2 Two Shapes of Failure and Infant Mortality Curves

	Infant mortality	Random failure (overstress failure)	Wear-out failure (degradation failure)
Core concept	Improper design and poor quality control	Two types of typical failure in item	
Cause of failure	Frequent encounter of great stress and damaged material	Accidental release of damaged material or a large stress on good material	Material wear or degradation near lifetime
The form of failure rate	High failure rate at first, decreasing thereafter	Constant failure rate	Increasing failure rate near lifetime
Probability density function	Weibull distribution function ($\beta < 1$)	Exponential distribution function (Weibull distribution function when $\beta = 1$)	Weibull distribution function ($\beta > 1$)
Application	Three shapes form bathtub curve; not applicable		
	Excluded completely	Two shapes form hockey stick line; applicable	

has no infant mortality, which can be precipitated by a short period testing and should consequently always be corrected.

In random failure, the cumulative failure rate is proportional to the lifetime acceptable to customers. Common sense tells us that an item with a 1% failure rate for 1 year has a 10% failure rate in 10 years. In a practical sense, this proportionality applies at below about 20% of the cumulative failure rate, which is the core concept of the exponential distribution function.* No proportionality during use indicates that wear-out failure and infant mortality are mixed with random failure, which presents numerous problems.

Let's consider the Weibull distribution function ($\beta > 1$) with regard to wear-out failure. It is important to understand what the magnitude of the shape parameter indicates. The higher the shape parameter, the more

---

* The failure rate of the exponential distribution function is approximately as follows when the product of the failure rate multiplied by time, ($\lambda \times t$), is sufficiently small: $F(t) = 1 - R(t) = 1 - e^{-\lambda \cdot t} \cong \lambda \cdot t$. The cumulative failure rate, $F(t)$, is proportional to time due to a constant failure rate. As the failure rate is defined as the ratio of failed numbers over total operating numbers and the total number of operating devices decreases due to failure, the cumulative failure rates calculated by approximation are greater than those acquired using the full equation; 10% by approximation is actually 9.52%, 20% is 18.13%. This is a reduction of 0.48% points and 1.87% points, or 4.8% and 9.4%, respectively.

intensively failure occurs, whether or not the same wear-out failure mechanism is involved. This type of curve, with a high shape parameter, indicates that there will be many failures within a short period of time as the item approaches its lifetime. Therefore, the larger the shape parameter, the more its failure rate will increase near lifetime. The shape parameter can thus be regarded as reflecting the intensity of the wear-out failure. Sometimes, wear-out failure due to the same failure mechanism occurs during use before lifetime. In this case, the smaller the shape parameter, the more frequent the failure. A large shape parameter indicates a reduced occurrence of wear-out failure within the lifetime. Therefore, the bigger the shape parameter, the better.[*] Obviously, it is desirable to have no item failures until the expected life, although of course many failures will arise after the anticipated lifetime.

As experience teaches, the lifetime of items that have virtually the same or very similar structures varies according to operational conditions, but the shape parameter itself is invariable.[†] This becomes the basis for accelerated life testing of the item.[‡] The advantage that the shape parameter can be extracted from past data across failure mechanisms enables us to apply the Weibull distribution function to designing and analyzing life testing for wear-out failure.[§] But since good design—robust structure and homogeneous material—focuses failures so that they occur at a certain time, the shape parameter will rise.

In an item with a low level of reliability, wear-out failures accumulate and occur, along with random failure, within the projected lifetime. In this case, lifetime should be calculated only in terms of wear-out failure because it is irrational to reduce the expected lifetime because of random failures.[¶] Meanwhile, the failure rate within the lifetime is estimated considering all failures, because it is risky for an item to have a high failure rate within its lifetime. Wear-out failures should be analyzed through the Weibull distribution function, but failures within the lifetime ought to be calculated with the exponential distribution function, even if some of them are not random failures.[**] When undertaking corrective action to

---

[*] This is explained in Chapter 3, Section 3.6.

[†] R. Fujimoto, "Lifetime Estimation by Weibull Analysis (Excel)," *Proceedings of the 30th Reliability and Maintainability Symposium*, Union of Japanese Scientists and Engineers, 2000, p. 246.

[‡] See Chapter 5, Section 5.5.

[§] Sometimes the lognormal distribution function fits the data well. But presuming its parameter is difficult and requires additional data obtained from testing similar structures, it is convenient to apply the Weibull distribution function and complement it later.

[¶] This is discussed in detail in Chapter 3, Section 3.5.

[**] If failure within the lifetime includes wear-out failure, the failure rate would be comparatively high and the failure site would therefore be easy to identify. Without including wear-out failures, the failure rate simply reflects random failure. Sometimes, infrequent cases of wear-out failure within the lifetime cannot be identified with short periods of testing and must be found by life testing.

improve reliability, failures should be separated into categories of random failure and wear-out failure, and corrective actions should be designed correspondingly. Because wear-out failures sometimes arise during use before the end of the expected lifetime, there is the possibility they could necessitate an item recall. Generally, the failure rate can be reduced by investigating environmental stresses (one element of failure), and wear-out failure can be delayed by focusing on sites where the materials are weakest in the item structure (the other element).

3.3   Reliability target setting for finished products and unit items

Very unreliable items experience many infant mortalities, frequent random failures, and various kinds of wear-out failure throughout their lifetimes. These malfunctions can be decreased through design changes in the item structure, the thoroughness of manufacturing specifications, and a quality assurance system that analyzes initial failures, overstress failures, and wear-out failures. These steps should ensure little or no trouble until near the end of the item's lifetime, when the failure rate inevitably increases. In this case, the shape of the failure rate over time is similar to a hockey stick, or the bathtub curve without its left wall, as shown in lower line III in Figure 3.1. An item with a bathtub curve failure rate should not be released into the market—that is, infant mortality should be completely eliminated first. In actuality, the failure rates of nearly all highly reliable finished products and their components follow a hockey stick line.

At this stage, decisions must be made about how low failure rates should be and how long the item lifetime will be. These will be explained separately for completed products for customer use and for the units incorporated into them.

Generally, the expected lifetimes of finished products in use are acknowledged by customers. For example, home electrical appliances ought to last about 10 years. The lifetime of a mobile phone should be over 5 years, and if a competition between primary performance and additional features is set up, it could be over 10 years. The lifetimes of durables are surveyed and reported once a year in the daily newspaper *Dempa-shimbun* (Japan Electric Wave Newspaper)[*] and the monthly magazine *Consumer Reports* in the United States. For instance, *Dempa-shimbun* reports that, on average in Japan, new cars are used by the primary purchaser 6 years before passing into the used car market, and that secondhand cars are used another

[*] D. Ryu, *Technology Revolution*, Paju, Korea: Hanseung, 1999, p. 100.

6 years; the average car is thus driven a total of 12 years and scrapped thereafter. This schedule is in line with general customer expectations.

Now let's consider the basic concepts for targeting failure rates within lifetimes for assembled products. The technological complexity of finished products does not excuse their failure rates rising. Customers regard the product as a modern convenience expected to give good performance for a reasonable time. Thus, the average annual failure rates of finished products should be set below a certain level, regardless of the type of product, and would decrease gradually as technology advances. Although not many items are in this category, the failure rates of durables are surveyed and reported in *Consumer Reports* mainly as after-service rates for 5 years. This rate, divided by 5 (years), becomes the average annual service rate; twice the annual service rate is the annual failure rate.[*] The target failure rate over the lifetime of an assembled product should be low enough that customer complaints and Internet postings about repairs or failures cannot negatively influence the brand of the product or company.[†] Reducing the volume of complaints can result in excessive service expenses, which will shrink profits. The target must be set considering these factors. The annual failure rate of home electrical appliances is 1 to 3%, and their cumulative failure rates reach 10 to 30% over 10 years.[‡] The annual failure rate for automobiles is 3 to 4%, and the cumulative failure rate reaches 36 to 48% after 12 years; this would be decreased to a level closer to that of electric appliances. Total service expenses will then be around 1% of sales revenue or less.

So much for the reliability targets for assembled products. How then can the reliability targets for unit items incorporated into these products be reasonably established?

Let us assume that the number of components in unit items is about 100. First, the lifetime target of each unit should be equal to or longer than that of the completed product, because the lifetime of the assembly will be determined by the life of the shortest-lived unit in it. (The lifetime target will be discussed in the next section.)

The annual failure rate of a component is calculated from the failure rate of the assembled product. If the failure rates of all units follow a hockey stick line, the sum of the failure rates of all units comprising the finished product will become the failure rate of the finished product. For example, if the target annual failure rate is 1% and the number of major units reaches 10, then the target of the annual failure rates of each unit will be around 0.1%. Although unit rates cannot actually be averaged due to the technological disparities among them, in the long run the sum of all

[*] Ibid., p. 81.
[†] Ibid., p. 96.
[‡] See Chapter 2, Section 2.2.

Table 3.3 Targets for Unit Items for Various Products

	Refrigerator	Motor	Airplane
Annual failure rate	1%/year	1%/year	1%/year[a]
Number of components[b]	1,000	10,000	100,000
Number of units (about 100 components)	10	100	1,000
Annual failure rate of unit	0.1%/year	0.01%/year	0.001%/year
Annual operation time	2,000 h	2,000 h	2,000 h
Hourly failure rate of unit	5×10^{-7}/h	0.5×10^{-7}/h	0.05×10^{-7}/h

[a] Exclude catastrophic failure, such as in a nosedive crash.
[b] The number of components is reduced for easy comparison. The number of components is actually about 2,000 in the refrigerator, 20,000 in the motor, and 700,000 in a large-sized airplane.

the unit rates maintains the overall rate at 1% or below. Table 3.3 illustrates this concept.

The target annual failure rate for any given unit varies widely, depending on the final product in which it is incorporated. For example, as shown in Table 3.3, the annual failure rate of a cathode ray tube may be satisfactory at about 0.1% for a home-use television, but its failure rate in an airplane instrument must be as low as 0.001%, even without considering the more severe environmental conditions the plane encounters. The more unit items are incorporated into the final product, the lower, in inverse proportion, its failure rate should be. Under these circumstances, it is difficult to meet the reliability targets for complex products, and attaining the target failure rate is no easier than attaining the target item life. This is because the target is very low, and lowering it further is difficult.[*] Likewise, since the high-quality version of a given product type has more units or components, its targeted failure rate ought to be comparably adjusted, relative to a lower-end product.

The same concept should apply to components under development with increasing complexity, such as microcircuit devices for use in computers. For example, computer performance has been improved by the advancement of the microchip, but the failure rate of the computer should stay constant or be lower for each new chip generation. Moore's law says that the density of a microchip doubles every 18 months, and thus chip performance and computer performance should be twice as efficient. Since with each generation there are roughly twice as many elements on the same size chip, its failure rate will increase by two times or more, which would result in poor reliability and customer dissatisfaction with the finished computer. For actual use, the failure rate should be reduced by half or more, until the failure rate of the new computer is the same or better than that of

[*] This is explained in Chapter 5, Section 5.7.

the current model. Meanwhile, according to the verified results,[*] Pecht's law states that the failure rates of microcircuit devices for digital systems decrease by about 50% every 15 months. Therefore, the duration of the relevant period in Moore's law cannot be shorter than that of Pecht's law.[†]

3.4 The BX life target setting as a durability index

Using the mean time to failure (MTTF) as a lifetime index leads to misunderstanding, whereas the index of BX life more properly reflects the real scenario of the life of an assembled product.[‡] The last section explained that the lifetime target of a unit incorporated into a more complex product should be equal to or longer than that of the completed product. Moreover, the lifetime should be considered as the point at which the failure rate rapidly increases due to wear-out failure. The lifetime failure rate for individual units incorporated into a complex product should be very low.

Now, consider what the lifetime of each unit is. We have repeatedly said that item lifetime is the time when wear-out failure begins. But if the cumulative failure rate, including random failure, is quite large before the point at which the unit items wear out, this point cannot be considered the lifetime. If the failure rate of the individual units is high before the lifetime is reached, the failure rate of the final product that includes these items will also be high, and therefore the final product will have little value. For example, a television with a 10-year lifetime (B10 life) cannot incorporate units having a 1% annual failure rate for 10 years. Then each unit would cause a 10% failure rate during the 10-year lifetime and all the televisions comprising 10 units will fail in 10 years. Thus, the lifetime of the component unit should be reduced to the time when it reaches a certain level of the cumulative failure rate in advance of the rapid failure rate increase when intensive failures occur due to the wear-out of the unit. In other words, all failures within the lifetime, including overstress failure, should be below a certain level before wear-out failure of the item occurs intensively. Thus, the BX life becomes the time when a certain limit (X) of the cumulative failure rate is reached (although the setting is not easy to determine). Thus, we use BX life as a durability index that is applied to both the assembled product and its components.[§] So both the assembled product and its units have the same lifetime (BX life). But the failure rates (X of BX life) of the units must be much lower than that of the assembled product because the sum of all unit failure rates must be the same or smaller than that of the assembled product.

[*] See Chapter 2, Section 2.5.
[†] Chapter 5, Section 5.7, describes the necessary procedure to shorten the period of Pecht's law.
[‡] See Chapter 2, Section 2.2.
[§] See Chapter 2, Section 2.2, on the lifetime of the completed product.

How low is the limit (X) of the cumulative failure rate expressed in BX life? Compare the two rates—the cumulative failure rate of each unit item before the intensive occurrence of wear-out failure and the cumulative failure rate of units allocated when included in the finished product. If the former is greater than the latter, the lifetime of the unit item should be reduced to the point when the allocated failure rate is reached. Therefore, the lifetime of the unit varies according to the finished product incorporating it.

Assume that the cumulative failure rate of a unit having about 100 components is 1.2% for 12 years, or that the B1.2 life is 12 years, which means B1 life 10 years and B0.1 life 1 year, limited only by overstress failure; also assume that from the 13th year, it increases over 10%, overlapping with wear-out failure. Let's estimate the BX life of this unit when it is incorporated into both a refrigerator and a motor. Since there are 10 units like this in the refrigerator and 100 units in the motor, and the cumulative failure rates of the finished product are the same (10% for 10 years, or B10 life 10 years), the required cumulative failure rate of the unit incorporated in the refrigerator and the motor will be 1 and 0.1%, or B1 life 10 years and B0.1 life 10 years, respectively. So this unit can be reliably incorporated into the refrigerator, because the lifetime of the refrigerator, having 10 units at a 10% failure rate, will be 10 years, or B10 life 10 years. But if it is incorporated into the motor, its lifetime will be reduced to 1 year, or B0.1 life 1 year, because the motor has 100 units, which means each unit will have less than a 0.1% cumulative failure rate. Therefore this unit could be used effectively in the refrigerator but could not in the motor. In order for it to be used in the motor, the cumulative failure rate would have to be reduced from 1% for 10 years to 0.1% for 10 years, which means the B0.1 life should be increased from 1 to 10 years. After this improvement, the cumulative failure rate of the motor would be 10% for 10 years, or B10 life 10 years, because the lifetime of each of the 100 units is B0.1 life 10 years, as shown in Table 3.4.

Table 3.4 Target (X%) Applicability of a Unit When Incorporated into More Complex Product

		Refrigerator	Motor	
Unit used in final product	BX life	B1 life 10 years	B1 life 10 years	B0.1 life 10 years
	Cumulative failure rate for 10 years	1%	1%	0.1%
Final product	Number of units	10	100	100
	Cumulative failure rate for 10 years	10%	100%	10%
Applicable		Yes	No	Yes

The instance in Table 3.4 is related to overstress failure, but it would be the same in the case of wear-out failure without overstress failure. Assume that a certain unit item has not experienced overstress failure but only wear-out failure, and that the cumulative failure rate is 1% for 10 years, or B1 life 10 years. This unit can be used for the refrigerator but cannot also be used in the motor. If the shape parameter of the wear-out failure is 2, the B1 life 10 years of this unit is equivalent to B0.1 life 3.2 years, which needs to be extended to 10 years for the motor, or B0.1 life 10 years.[*] Because the X of the BX life of a unit or component, or the target cumulative failure rate for the lifetime of the final product, varies according to the final product, the BX life obtained from the test results should be converted into the changed X, which is allocated from the cumulative failure rate of the completed product over its lifetime. Thus, it is important to carefully estimate the item life.[†]

As mentioned earlier, although the final product may be complex and utilize advanced technology, it is not acceptable for it to have a high cumulative failure rate. The target lifetime of the unit items incorporated into complex products should be changed to, for example, B1 life Y years for a refrigerator and B0.1 life Y years for a motor. Therefore, in order to incorporate the unit item into an airplane—assuming the number of components in the airplane is 100,000, or the number of units (100 components equals 1 unit) is 1,000—the level should be even further reduced to B0.01 life Y years. *The New Weibull Handbook*[‡] indicates that the aerospace industry would apply a B1 life for benign failure, a B0.1 life for serious failure, and a B0.01 life for catastrophic failure, but this is inadequate. Table 3.5 summarizes this concept.

In the case of components like bearings, it is useful to compare each with the B10 life, but this cannot be done in terms of the lifetime itself. For example, assume that a certain item has 10 bearings, for which the lifetime is B10 life 3 years. This means that after 3 years, 100% of the items will fail. Since a bearing is not a unit but a component, it is desirable to convert the B10 life into at least B0.01 life or B0.001 life to correspond with the final product in which it is included.

Although the wear-out failures of similar items are epidemic after 10 years, additional effort should be devoted to improving an item with

---

[*] From Equation (3.4),

$$\left(\frac{L_{BX_1}}{L_{BX_{0.1}}}\right)^2 \cong \frac{1\%}{0.1\%} = 10, \left(\frac{10}{L_{BX_{0.1}}}\right)^2 \cong 10 \cdot \therefore L_{BX_{0.1}} \cong \frac{10}{\sqrt{10}} = 3.17 \text{ years.}$$

[†] See Chapter 3, Section 3.6, about the conversion of the BX life.

[‡] R. Abernathy, *The New Weibull Handbook*, 5th ed., North Palm Beach, FL: Abernethy, 2004, p. 2-6.

Table 3.5 Target of X in the BX Life for a Unit with 100 Components Varies according to the Final Product

		Refrigerator	Motor	Airplane
Final product	BX life[a]	B10 life 10 years	B10 life 10 years	B10 life 10 years
	Cumulative failure rate for 10 years[b]	10%	10%	10%
	Number of components[c]	1,000	10,000	100,000
	Number of units	10	100	1,000
Unit used for final product	Cumulative failure rate for 10 years	1%	0.1%	0.01%
	BX life of unit	B1 life 10 years	B0.1 life 10 years	B0.01 life 10 years

[a] All were used for over ten years, but their lives are reduced to ten years for easy explanation.
[b] Exclude catastrophic failure, such as would occur in a nosedive crash.
[c] The number of components has been reduced to simplify the example. The actual number of components is about 2,000 in the refrigerator, 20,000 in the motor, and 700,000 in a large airplane.

Table 3.6 Comparison of Lifetime Indices of a Unit Item

	MTTF	Life of wear-out failure start	BX life
Concept	When the cumulative failure rate is about 60%	The inflection point when the failure rate is increasing rapidly	Cumulative failure rate is X%
Remark	Does not fit customer's expectation	Meaningless if cumulative failures exceed a certain level before this time	X varies with the finished product and therefore item lifetime varies accordingly
Applicable	No	No	Yes

a 0.1% cumulative failure rate for 10 years rather than an item with a 1% cumulative failure rate for 10 years.[*]

As we have established, the lifetime is not the point in time at which the failure rate has rapidly increased due to wear-out failure, but the point at which the cumulative failure rate reaches a percentage distributed for the final assembled product. Therefore, BX life as a durability index adequately describes item life in real field circumstances, and the X of the BX life varies according to the finished product. Table 3.6 summarizes this concept.

[*] This will be discussed in Chapter 5, Section 5.7.

3.5 The revised definitions of failure rate and characteristic life

From the definition of reliability, we have deduced that the reliability issues to be confirmed are lifetime and failure rate within lifetime, and that the desirable graph of the failure rate follows a hockey stick line. In other words, the failure rate of an item within its lifetime is positioned near the X axis, or time axis, and increases with the life of the item. We also clarified that as a durability index, BX life is adequate, and that the X of BX life varies with the finished product.

Since the hockey stick line represents the ideal case, let's consider what is going on if the failure rate of an item does not follow this pattern. Recall that failure data include infant mortality, random failure, and wear-out failure. Based on these data, the failure rate of the item can be predicted correctly but the lifetime will be reduced in actual practice. Because the item lifetime will be longer if initial failures are reduced, finding and fixing initial failures is not difficult, and there is no justification for allowing initial failures to influence the lifetime. It has also been mentioned that the inverse of the MTTF calculated from random failures should not be considered as the lifetime, because random failure cannot occur indefinitely or be anticipated to be in the neighborhood of the lifetime calculated solely by considering wear-out failure. Thus, lifetime should be estimated based only on wear-out failure. The item lifetime should be the same both before and after improving infant mortality and accounting for random failures.

As an analogy, consider human beings. Because nowadays children do not die from smallpox or tuberculosis, a child who avoids serious disease for about a year has a long life expectancy of about 70 years.* Thirty years ago, life expectancy was over 60 years; if the influence of infant diseases is discounted, the average human lifetime then would have been around 70 years, or the same as that of today. Even the Bible said, maybe 3,200 years ago, that the length of our days is 70 years, or 80, if we have the strength.† In prehistory, a man 100 years old was a rarity, but not so today; modern medical science has lengthened human life expectancy by eliminating childhood diseases. There is excellent recent research on viruses, which are equivalent to overstress failures, and a little progress on treating degenerative diseases, which are analogous to wear-out failure. To really add to life extension, the index of life expectancy should be reconsidered and announced publicly, which would boost research into degenerative diseases.

---

* See Table 2.1.
† Psalm 90.

In sum, because the lifetime of hardware is defined as the time at which wear-out failure occurs intensively, it should be estimated only by considering wear-out failures. Thus, the kinds of failure for a given item should be analyzed, and random failure should be excluded in calculating the lifetime. Note that wear-out failure mechanisms related to lifetime are different from random failure mechanisms. Likewise, it is necessary to distinguish the two kinds of failure mechanisms when establishing test specifications, as well as when analyzing failure data. Short periods of testing under normal conditions may not precipitate wear-out failure and therefore cannot be used to accurately estimate lifetime because of the limitations of test specifications, as explained in the last paragraph of this section.

In products we now accept as modern conveniences, there is no infant mortality, only rare random failures under particular environmental conditions, and wear-out failures at a certain site after the item life is reached. In unit items incorporated into complex durable products such as motors, we should reduce both wear-out failure during use and random failure. For example, the cumulative failure rate of an item used in home appliances should be reduced from 1% to 0.1% over its lifetime if it is to be used for a motor, and to 0.01% for a more complex use, such as in an airplane. This, or delaying wear-out failure, needs to be done after reducing infant mortality and random failure. We frequently call this *reliability growth*, which is an inaccurate term, since it includes the failure rate; a more exact expression would be the *durability growth* or *life extension*.

Now consider the equation defining lifetime and failure rate. The lifetime is defined as the total test hours, or the test hours multiplied by the sample number, divided by the failed number. Likewise, the failure rate is defined as the inverse of the lifetime, or as the failed number divided by the total test hours:

Lifetime   = (total test hours)/(failed number) = (test hours) × (sample number)/(failed number)
Failure rate = (failed number)/(total test hours)

Let's examine the adequacy of these equations. Assume that 100 units of an item have been tested for 1,000 h, or have been time censored. When there is no failure, the failure rate will be 0% per hour. Then assume that 1,000 units of the same item have been tested for 1,000 h. If there is still no failure, the failure rate will also be 0% per hour. As mentioned with regard to estimating the defect rate,[*] it is unreasonable to express the results as the same figure in spite of different test executions. Finally, let's estimate the unit lifetime. Since there is no failure, we cannot calculate it because the denominator is zero. This definition may have been

* See Chapter 2, Section 2.3.

applicable to items in earlier times, when the level of reliability was low, but cannot be adapted to current items of high reliability, for which it is difficult to easily document failures. It is thus necessary to reshape the above definitions.

The contradiction can be resolved by using a minimum value in estimating the lifetime and a maximum value in estimating the failure rate and assigning a confidence level. As mentioned earlier, applying the "add 1 rule" (add 1 to the failed number) to the term *failed number* in the definition achieves a good solution with a commonsense level of confidence of around 60%. Calculated this way, the failure rates in the above cases would be 1% per 1,000 h and 0.1% per 1,000 h, respectively, and the lifetimes would be 100,000 h and 1 million h, respectively (which are the inverse of the failure rates), where the shape parameters are 1.*

Therefore, I propose a revised definition of failure rate (λ)[†] and lifetime (η)[‡] as follows, where the confidence level is around 60%. Lifetime in this case is called *characteristic life*, which should be calculated only from wear-out failure. Time variables are adjusted into the shape parameter exponent of the time variable, or the intensity of wear-out failure, β. Thus,

$$\lambda \cong \frac{r+1}{n \cdot h} \tag{3.1}$$

$$\eta^{\beta} \cong \frac{n \cdot h^{\beta}}{r+1} \tag{3.2}$$

where r is the failed number, n is the sample number, h is the test hours, and β is the shape parameter.

In the case of 100 units tested for 1,000 h, the failure rates and lifetimes estimated by the current definition and the revised definition based on the failed number are shown in Table 3.7. Comparing the two results, we can conclude that the confidence level of the current definition is far lower than

---

* Converting the data of random failures into lifetime is not reasonable, but is used for ease of understanding.
† H. Shiomi, *Reliability Practice*, Union of Japanese Scientists and Engineers, 1977, p. 163.

$$\lambda = \frac{\chi_{\alpha}^{2}(2r+2)}{2} \cdot \frac{1}{\sum t_i} \cong \frac{\chi_{\alpha}^{2}(2r+2)}{2} \cdot \frac{1}{n \cdot h},$$

where, substituting $\alpha = (1 - 0.63)$, $r = 0$, $\alpha = (1 - 0.59)$, $r = 1$, and $\alpha = (1 - 0.58)$, $r = 2$. Then the first term equates to $(r + 1)$, or 1, 2, 3.

‡ See Chapter 5, Section 5.6, footnote asterisk (p. 104) and R. Abernethy, *The New Weibull Handbook*, 5th ed., North Palm Beach, FL: Abernethy, 2004, pp. 6-2, 6-4. In case that $r \geq 1$.

$$\eta^{\beta} = \frac{2}{\chi_{\alpha}^{2}(2r+2)} \cdot \sum t_i^{\beta} \cong \frac{2}{\chi_{\alpha}^{2}(2r+2)} \cdot n \cdot h^{\beta}$$

Table 3.7 Comparison of the Current Definition and the Proposed Definition with a Commonsense Level of Confidence (100 Units Test for 1,000 h)

Failed Number	Failure Rate (λ)		Characteristic Life (η; $\beta = 1$) (2)		Characteristic Life (η; $\beta = 2$)	
	Current definition	Proposed definition	Current definition	Proposed definition	Current definition	Proposed definition
0	0	1%/1,000 h	Incalculable	100,000 h	Incalculable	10,000 h
1	1%/1,000 h	2%/1,000 h	100,000 h	50,000 h	10,000 h	7,070 h
2	2%/1,000 h	3%/1,000 h	50,000 h	33,300 h	7,070 h	5,770 h

that of the proposed definition. When there is one failure, the confidence level of the prediction "1%/1,000 h" will be 26%, and that of the prediction "2%/1,000 h" will be 59%, or the commonsense level of confidence.[*]

Let's examine the meaning of the revised definitions, Equations (3.1) and (3.2). If the test conditions, or sample size and test period, are sufficient to identify failure, or actually find at least one, the results are feasible. But if the conditions are insufficient, there will be no failure, and therefore the failure rate will be estimated as a larger value and the lifetime as a shorter value. For example, assume that there is an item with a failure rate of around 1% per 1,000 h. If we test 100 units of the item for 1,000 h and find no failure, the failure rate becomes 1% per 1,000 h, according to the revised definition, and then there is no difference from the actual rate. Meanwhile, if we test 100 units of the same item for 500 h, we are destined to find no failure and the failure rate becomes 1% per 500 h, or 2% by 1,000 h by the revised definition, which is an incorrect estimation, or two times greater than the actual failure rate. Thus, in order to estimate properly, test conditions should approach as nearly as possible the conditions under which failures actually occur.

When the test conditions are insufficient, what does the revised definition mean? The results produced by such test conditions express the limitations of the test conditions themselves. As mentioned earlier, the desirable values for a lifetime or a failure rate are already determined by the requirements of the customers and the market situation. It is therefore possible to figure out whether the designed test conditions are sufficient to confirm the established goals or not. Thus, the revised definitions constitute the basic equations to check the validity of test conditions with regard to target fits. Many test specifications are inappropriate for making reliability judgments.[†] In other words, many specifications have below half confidence, or 50%, confidence levels.

[*] See Table 2.4.
[†] See Chapter 5.

3.6 The conversion from characteristic life to BX life

The concept of failure rate is the arithmetic mean of failure rates in a product lot, and so is applicable to actual fields. Characteristic life, when the shape parameter of an item is 1, is equal to the MTTF, which creates misunderstanding.[*] Likewise, the characteristic life (η) is similar to the MTTF, to which it has a different proximity according to the shape parameter. Specifically, it is equal to the period from the start of item usage to the time when 63% of the items of one production lot fail.[†] But it is better to use the BX life as the lifetime index. Therefore, the characteristic life should be converted into the BX life, or the time when the accumulated failure rate reaches X percent. The relation between the characteristic life (η) and the BX life (L_{BX}) is[‡]

$$L_{BX}{}^{\beta} \cong x \cdot \eta^{\beta} = \frac{X}{100} \cdot \eta^{\beta} \qquad (3.3)$$

where $X = 100 \cdot x$ and X is the cumulative failure rate (%).

This equation is only an approximation when X percent is small. It is convenient to apply because the X of the BX life of the unit item would be below 10%, and thus the difference between the full equation and the approximate equation will be only a few percent.[§] The BX life diminishes as much as X percent relative to the characteristic life, but the greater the shape parameter, the less the diminishment will be. For reference, in the case of random failure, the BX life is equal to X percent of the characteristic life—a conversion that is irrational, since there is no relation between random failure and item life, as we have established.

There needs to be a conversion between BX lives as one X changes to another X. For example, the B10 life of a bearing should be converted to the

---

[*] See Chapter 2, Section 2.2.

[†] The cumulative failure rate of the Weibull distribution function is as follows. $F(t) = 1 - R(t)$ $= 1 - e^{-(t/\eta)^{\beta}}$, where if $t = \eta$, $F(\eta) = 1 - e^{-1} = 0.63$.

[‡] As BX life is the period when the cumulative failure rate reaches x, reliability at this time is $(1 - x)$. That means $R(L_{BX}) = e^{-(L_{BX}/\eta)^{\beta}} = 1 - x$,

$$\therefore L_{BX}{}^{\beta} = \ln(1-x)^{-1} \cdot \eta^{\beta} = \left(x + \frac{x^2}{2} + \frac{x^3}{3} + \cdots \right) \cdot \eta^{\beta},$$

where the cumulative failure rate (x) is sufficiently small. Then $\ln(1-x)^{-1} \cong x$. Finally, we get $L_{BX}{}^{\beta} \cong x \cdot \eta^{\beta}$.

[§] As shown in footnote double cross on this page, the error due to x is as follows: when the shape parameter is 2, the B1 life decreases from 0.10025 η to 0.100 η, the B5 life from 0.226 η to 0.224 η, and the B10 life from 0.325 η to 0.316 η, which reduces by 0.2, 1.2, and 2.6%, respectively.

B0.01 life or B0.001 life, matched to the completed product incorporating the bearing. The conversion equation can be derived from Equation (3.3) as follows (the shape parameter of the bearing is 2~3):[*]

$$\frac{L_{BX_2}{}^\beta}{L_{BX_1}{}^\beta} \cong \frac{x_2}{x_1} = \frac{X_2}{X_1} \tag{3.4}$$

where L_{BX_1}, L_{BX_2} are the BX lives at failure rates $X_1\%$ (= $100 \cdot x_1$), $X_2\%$ (= $100 \cdot x_2$), respectively.

Table 3.8 shows characteristic lives calculated first by revised Equation (3.2), when there are no failures under four test conditions, and the B10 lives, B1 lives, and B0.1 lives are converted by Equation (3.3). Notice that the decrease in characteristic lives to B0.1 lives has a constant rate according to the shape parameter; that is, the rate decreases from characteristic lives to B10 lives, from B10 lives to B1 lives, and from B1 lives to B0.1 lives.

This is important. In order to obtain a proven BX life, the BX life should be equal to or less than the test hours and the proven BX life should also be converted to a smaller X. Take a look at Table 3.8. Test numbers 1 through 6 were conducted for 1,000 h and confirmed that trouble occurred during that period. In Table 3.8, characteristic lives or BX lives in parentheses are over 1,000 h. Thus, lives equal to or less than 1,000 h are proven results, which are the B10, B1, and B0.1 lives in test numbers 1, 2, and 3, respectively, and the B1 and B0.1 lives in tests 4, 5, and 6. A life over 1,000 h is unrealistic, due to only the arithmetic, because we did not allow that any failure might occur *after* 1,000 h. This is like regarding a town as a paradise of long life, based on the false prediction that fewer old men would die because of the rare deaths of youngsters. Therefore, the results in parentheses in Table 3.8 cannot be adopted. As far as lifetime testing is concerned, lives that can be determined within the test hours are proven results, but conversions beyond test hours are limited.

Within the test period, the failure occurrence retards with the magnitude of the shape parameter. Although the B10 lives are the same in 1,000 h for test numbers 1, 2, and 3, the B1 life and B0.1 life of large shape parameters are extended longer than those of smaller shape parameters. Unit items used for complex products like airplanes should be improved by trying to give them high shape parameters so they can be failed intensively.

The full probability function is not applied to reliability fields—only the initial portion is. Reliability statistics when the cumulative failure rate is small still need to be researched and developed.

Let's mentally calculate the BX lives from the test conditions. We can assume that when no failure occurs in 10 units during the test, it indicates

---

[*] In addition, when two or more failures are found, the magnitude of the shape parameter can be estimated using Equation (3.4).

Table 3.8 Characteristic Life, B10 Life, B1 Life, and B0.1 Life for a Commonsense Level of Confidence

Test number	Test condition	Failed number	Shape parameter	Characteristic life	B10 life	B1 life	B0.1 life
1	10 units—1,000h	0	1[a]	(10,000)	1,000	100	10
2	10 units—1,000h	0	2	(3,160)	1,000	316	100
3	10 units—1,000h	0	3	(2,150)	1,000	464	215
4	100 units—1,000h	0	1[a]	(100,000)	(10,000)	1,000	100
5	100 units—1,000h	0	2	(10,000)	(3,160)	1,000	316
6	100 units—1,000h	0	3	(4,640)	(2,150)	1,000	464
7	25 units—2,000 h	0	2	(10,000)	(3,160)	1,000	316
8	12 units—2,000 h	0	3	(4,580)	2,130	986	458

[a] Converting the data of random failures, of which the shape parameter is one, into lifetime is not reasonable, but is simplified here for demonstration purposes.

a 10% cumulative failure rate, and no failure in 100 units tested indicates a 1% cumulative failure rate for 1,000 h, with a commonsense level of confidence. Thus, we can say that the conditions for test numbers 1, 2, and 3 confirm B10 lives 1,000 h, and the conditions of test numbers 4, 5, and 6 confirm B1 lives 1,000 h. It is not necessary to calculate the characteristic life first and then convert it into the BX life.* A clear understanding of the concepts in Chapter 5 should lead to a mental calculation of the reduced sample size needed to attain the same reliability target, such as 25 units for test number 7, as opposed to 100 units for test number 5, or 12 units for test number 8, as opposed to 100 units for test number 6, when the test hours increase from 1,000 to 2,000.

3.7 Connotations of these new concepts†

Among the concepts clarified so far, several have important ramifications for the field of reliability technology. These include, particularly, the facts that the failure form cannot be differentiated between electronic items and mechanical items, since the two common elements of failure are stress and materials. It is also important to recognize that random failure and wear-out failure can coexist in one item, and that the BX life should be selected as the lifetime index. A final crucial fact is that the definitions of the two reliability indices should be adjusted. Many things remain to be done to advance reliability technology. Current reliability technology is built like a house on sand, not rock. Establishing new reliability concepts, relevant index selection, and index definition implies reconstructing the foundations of the field.

This makes it apparent why reliability accidents still frequently occur. These failures can be classified into two groups: failure due to uncontrollable stresses, such as natural disasters, and poor product design and manufacture regarding more typical stresses. These design flaws are supported by the untrustworthy foundations of current reliability technology. The corrective actions for natural disasters are prediction and bypassing

---

* The basis for this is discussed in Chapter 5, Section 5.6.

† These novel concepts of reliability technology and methodology, confirming the two reliability indices necessary for product verification, are based on over 30 years of field experience. In 2005, the author published a brief paper on these concepts in the *Journal of Microelectronics Reliability* ("Novel Concepts for Reliability Technology," 45(3–4), 611–622). In June 2006, he presented a 70 min plenary speech at the 36th Reliability and Maintainability Symposium organized by Union of Japanese Scientists and Engineers (JUSE) (also titled "Novel Concepts for Reliability Technology," *Proceedings of the 36th Reliability and Maintainability Symposium*, JUSE, pp. 21–41). The postsymposium report indicates that these reconceived foundational ideas are now being used for new development, and that back-to-basics reliability research and execution is now being implemented in Korea, while Japan is using a more "rational" (if not exactly scientific) approach to reliability technology. See http://www.juse.or.jp/reliability/symposium_36r&ms_repo.html.

stress (say, by evacuating the population), and therefore not relevant to product quality. Failures due to manageable stresses are inherent in products, and can and must be improved. Solutions for failures should not just be researched after they occur, but methodologies must be developed to find them in advance and to improve specifications confirming reliability before product release.

Next, we will consider the status of reliability technology and its embedded issues and extract the necessary direction for new methodologies based on the novel concepts of reliability technology presented so far.

chapter four

Current issues in reliability technology

Several years ago, a medium-sized company regarded as a rising star in Korea declared bankruptcy. It had developed mobile phones and launched them successfully in the domestic market. Due to heavy price competition in China, its profitability had deteriorated, so in the expectation of higher prices, it opened up a new market in Europe.* But there was also severe price competition there, and the company released its product without enough margin. This may have succeeded, but an inherent quality problem forced multitudes of product returns, which forced the business to close, finally, with major losses. Now, with the World Trade Organization (WTO) system developed from the General Agreement on Trades and Tariffs (GATT) in order to enhance free trade, the global marketplace has become increasingly competitive; companies can only survive when they make no mistakes in either price or quality.

It is dangerous for neophytes entering a market with few competitors to assume that high product sales will make it possible to match the existing market price, because the current strategy of price lowering by prevailing companies can easily collapse a recent arrival's business. Thus, newcomers should manage their businesses with small quantities of a product, avoiding being targeted by dominant suppliers until the business is on a safe foundation. To succeed, a new enterprise should be unique, with brilliant ideas for designing and manufacturing products and for managing business. But mistakes in the quality of the product can make a small business's losses irrecoverable. Why is this so?

There are two basic goals to be pursued in the business of making hardware products, as shown in Table 4.1. One is to meet customers' requirements; the other is to maintain an efficient business. Meeting customers' requirements means supplying good products. First, the product should fulfill basic quality standards, meaning that it performs at least as well as competing products, that it has no defects, and that it does not fail during use. But a product with just this level of basic quality does not attract customers' attention, and therefore can only be sold at a cheaper

---

* The market share cannot be increased by only advertising and public relations. The competitive advantage of superior product quality is also necessary. Despite substantial advertising, if a product is less competitive, word of mouth will label it as low grade.

Table 4.1 Business Activities Necessary to Remain Competitive

	Meeting customers' requirements	Maintaining an efficient business
Purpose	Customer satisfaction	Realizing surplus (creating profit)
Details	1. Basic quality (no defects or failures) 2. Good aesthetics and ease of use 3. Performance edge	1. Design for reduced material and manufacturing costs 2. Reduce development period and expenditures 3. Reduce marketing, production, and shipping expenditures 4. Conduct research and development for continuing competitiveness

price to ensure a profit. In order to gain customers, a product should also display good aesthetic design, ease of use, and special comparative features in performance.

Maintaining good business principles means making profits an ongoing concern—the result of four kinds of activities that give a manufacturer an edge over the competition. First, a product should be designed so as to reduce material and manufacturing costs. Second, the product should be developed over a shorter period and less expensively than those of competitors. Third, it should be produced, introduced, promoted, and conveyed to customers efficiently. Finally, various technologies relevant to the product should be studied and researched to ensure a continuing stream of high-quality future products.

Product design, the first of these features, has a great influence on realizing profit. Designing a product to be cheaper is not easy, but must be a constant goal in hardware manufacturing. The product should be designed to be as dense and compact as possible to meet customers' requirements. As the reduced material cost itself turns into profit, such a product already has a little margin built into the structure itself.

However, what if failures with similar failure modes occur epidemically after product release? It takes time to establish the needed corrective action because the exact cause must be researched and determined. Regardless of the cause, however, it is not easy to find an adequate corrective, because there is often not enough space on a densely designed structure to accommodate a remedy. It also takes considerable time to prepare spare parts and the repair site. One rapid response to failure is to substitute an identical new product for the failed one, but the same problem will occur again if it is not resolved. If lots of products have been released into a widespread market before the issue is identified and resolved, it is a very risky decision to commit all the resources of the company to repairing the returned products, and difficult to recover the damaged company image resulting from a recall.

Moreover, all this work is usually done primarily by performance designers, who are rarely specialists in reliability. This should make a CEO wary. In fact, few CEOs understand the significant differences between the field of reliability technology and that of performance technology. CEOs imagine a bright future in which their newly developed product will prevail throughout the market once it is released, but there may be an ambush waiting in the form of reliability failure. Distressingly, many promising venture companies, as well as large firms, have disappeared due to reliability problems 2 or 3 years after a promising product's release. Many companies have had to shut down a strategic business unit that had exhausted its surplus and laid off its personnel due to expensive reliability accidents. Because reliability concepts have been usually explained with complicated mathematics, these basic ideas have been difficult for many CEOs to grasp and apply.

Many companies confuse reliability technology with quality control activities. They think that checking performance includes checking reliability, or they regard a few days' operation of pilot products as a reliability test, which makes them think tests taking a few months are needless. The former is often the case with new venture companies that go bankrupt in 2 or 3 years after product release due to epidemic product failure. The latter is the case that service expenditures balloon after product release. CEOs require their employees to meet quality goals because this is basic to production, but often confuse reliability with conformance quality (c-quality), which can lead to costly, and even fatal, mistakes.

Why are reliability concepts so difficult to understand? To begin with, there are many methods or tests to be considered. Given the current state of the field, it is hard to judge which are the most effective, and also difficult to understand the relationships between the methods and the tests. Although it is well known that advanced companies use a variety of processes to confirm reliability, companies that establish their own tests to identify issues often are not convinced that their tests will generate good results. They may hesitate to investigate hidden issues, simply hoping that failures will not occur after the product is released into the market. Moreover, they are reluctant to introduce new reliability processes into their product development because to do so would require substantial changes in the business organization.

Let's examine this from the CEO's standpoint. CEOs question whether tests are really needed to confirm reliability issues, considering that the goal of engineering is usually to provide a design that does not require any actual tests. For example, the design of a mechanical structure, such as a bridge, is achieved by using mechanical principles, not tests. Of course it is better to check any product through testing, but many CEOs regard reliability tests as an additional expense, like life insurance. As the investment in test equipment mounts, CEOs doubt the wisdom of these expenditures, not realizing how cost-effective they may be. When product development is complete, some CEOs consider releasing the product in restricted quantities or in a predetermined test market, then watching the results over time. But these ideas are often no more than ideas—more often, the new product is sold aggressively to avoid losing competitive advantage. In the meantime, if epidemic failures occur, damages can snowball. Although it may be possible to keep the brand image after a major product failure, the risks of this strategy should be considered in advance. In my experience, the reality is that CEOs are unsure whether to believe practitioners' reports about product reliability. CEOs' technical knowledge is generally insufficient to allow them to capably check production activities in advance in order to ensure the reliability aspect of quality.

4.1   Product approval tests and their bases

Now let's consider the foundations for the specifications of product approval tests. First, the environmental and operational conditions that an item experiences from birth to discard must be assessed, and new products must be designed to work well within them. The circumstances regarding conveying the item from factory to the installed location must also be taken into consideration; it would be unfortunate if a carefully designed and manufactured item was broken during forwarding. Transportation vehicles—trucks, ships, and airplanes—all encounter unique and varied conditions. How the item is to be moved should be embodied in specifications as basic information. These conditions are classified into two groups: the conveying environments en route to customers, and the customer usage environments. Usage environments consist of the mechanical conditions imposed from outside, such as temperature and vibration, which produce stresses affecting the item, and the electromagnetic conditions under which the item gives and receives stresses due to the operational electricity used by products around it.

All too often it is assumed that the test methods applied in product approval are complex and should be left to reliability specialists, but they can be understood with common sense in the light of our experience. Let's review some of these reliability specifications, as shown in Table 4.2.

Let's consider the tests for conveyance environments. A drop test indicates what happens when an item falls—for example, as it is loaded or unloaded from the factory line or from the depot into a truck. The height of the test drop is set near the working position of the forklift or, in the case of, say, a mobile phone, at human chest height. Since a mobile phone

Table 4.2 Product Approval Tests by Conveyance and Usage Environments

	Conveyance environment	Usage environment	
		Mechanical environment	Electromagnetic environment
Test name	Drop, vibration, high temperature, high humidity, etc.	High temperature, low temperature, temperature cycling, high temperature plus humidity, vibration, shock, fatigue, etc.	Lightning surge, impulse noise, in-rush current, dip in power source, low power voltage, opposite power polarity, low battery voltage, electrostatic discharge, electromagnetic compatibility (EMC), fire, safety, etc.

is frequently dropped while being used (usage environment), it should be repeatedly tested without cushioning and packaging.

A vibration test is generated by synthesized vibrations that simulate various road conditions during conveyance. The item should set up for testing as it would be during shipping—protected with a shock-free cushion. If there is a cantilever type of component inside the item, it should be tightly tied up, because the shock cushion cannot prevent breakage caused by side-to-side swaying—just as a driver should fasten his safety belt, because the auto bumper cannot protect him from swaying back and forth.

If an item crosses the equator during shipping, testing it in high temperatures is necessary. When there is a large range of temperatures during the day, the moisture in the hot air around noon will be transformed into dew at dawn the next morning, which causes rust in materials like steel. Items shipped in these conditions require prior testing in high humidity. Items being shipped may also need a salt (brine) spray test; when applied to an item used outdoors, it becomes a test specification of the usage environment. To resolve moisture problems, moisture-laden air can be exhausted by vacuum packaging or absorbed by silica gel inside sealed packaging.

There are also several tests for the mechanical environments encountered in usage. If the item will be used in tropical or subtropical regions, it can be checked with a test for operation in high temperatures. Since increased temperature increases the energy in a material, which can transform it into another material altogether, heat inside the item should be exhausted cost-effectively by a fan or by natural convection. If the item is to be used in cold regions, it should be tested for operation in low temperatures, because at below-zero conditions materials often turn brittle.

Use in regions experiencing a broad range of temperatures requires tests that cycle from high to low temperature to repeatedly stretch and shrink the material to see if the materials degrade.

Testing for operation in high temperature and humidity checks the effects of moisture in hot, humid regions. Moisture reacting with contaminants causes corrosion in many materials, resulting in failed circuits. Alternatively, it can generate new microstructures, like dendrites, that can cause short circuits. Elevated temperature promotes this reaction, and voltage determines the direction of microstructure migration.

Since the conveyed product experiences the same vibration as the vehicle itself, tests of operation under vibration and shock conditions should be designed to fit anticipated road conditions. Repetitive stresses can cause many materials, including steel, to break suddenly after a certain period. This liability, called *fatigue*, can be confirmed with a fatigue test.

Product approval tests should be chosen based on the distinctive features of the environment where the item is to be installed, such as blowing sand or snowfall. For example, the item should be checked by tests that could suggest the need to improve filters to keep out sand or other substances, or to use materials that would be resistant against, say, the calcium chloride used to melt snow. Applicable tests of this sort can be easily selected by browsing through them.

Another important test evaluates the electromagnetic environment when a product is powered on. The test for lightning surge simulates the breakage or malfunction of components in a power circuit due to the high energy flow induced in the line by a thunderbolt. The phenomenon of lightning surge can be easily recognized; it resembles the flickering of a television screen. Atmospheric perturbation or the accumulation of charged solar particles forces the separation of positive and negative charges within the cloud. When the electric field becomes strong enough, an electrical discharge occurs within the cloud or between clouds and the ground. The high energy produced by the strike branches off into power lines or the signal line of electrical appliances. The resulting voltage surge, induced into a low-power line, may reach 20 kV. Around 85% of all strikes are below about 7 kV, as determined in Japan.[*] An electrical item should be able to endure several times the power of a triangle waveform with around 10 kV of voltage and hundreds of microseconds in duration.

Sparks sometimes discharge in the knife switches of equipment that consumes a lot of electric power. The power line from the transformer supplies electricity for many types of equipment in the area where a product operates. At the moment when electricity running into other appliances would be cut off, impulse noise due to these sparks may infiltrate

---

[*] E. Sugi, *The Design Manual for Lightning Surge Protection (TM-32)*, Osaka, Japan: Quality Department of Matsushita Electric, 1991, p. 19.

the product. Because the voltages of impulse noise are usually in the range of 2~3 kV, lower than that of lightning but with higher frequencies, the impulse passes through the power circuit of the item and throughout the microcircuits of digital devices, causing breakage or malfunctions. The item should be able to endure, for a few minutes, a square-waveform power surges of a few kilovolts lasting from fifty to a hundred nanoseconds, applied periodically for tens of milliseconds. The rated energy of the power surge varies with the product. When other nearby high-power-consumption equipment is consuming a great deal of electricity, big loads draw a lot of power, causing irregular power that skips a few hertz;* the item responds as it would to a stoppage of power. These cycles finally result in a malfunction. Testing for a dip in the power source accounts for such a phenomenon.

Another electrical problem can occur when a maximum voltage sine wave of electricity is conveyed to the item at the very moment when the item turns on. Current rushes to it suddenly and sometimes damages components in the power circuit, including fuses. The test of in-rushing current checks for such damages.

When all the air conditioners are operating in the summer, drawing a great deal of power, the voltages of the power sources decrease up to 20% of their rated output. The test for low power voltage evaluates this. Products to be used in countries with poor or questionable electric power supplies are particularly vulnerable to this problem.

There should be no problems with equipment using direct (not alternating) current when it is connected to battery terminals in reverse. This can be checked with the test of opposite polarity of power. Testing for low battery voltage is needed to check for abnormally low battery voltages approaching lifetime.

Static electricity produced by clothing, carpeting, equipment, and so on discharges through the hand and infiltrates the circuit, potentially causing overstress failure. Static electricity voltages are usually from a few to tens of kilovolts,[†] up to about 25 kV. Since the breakdown electric field strength in air is on the order of 30 kV/cm, electronic parts subject to such discharges should be separated by about 1 cm.[‡] The relevant test is the test of electrostatic discharge.

Nowadays various electric appliances are used together in small areas. Electromagnetic energy is transmitted through air or power cables, sometimes with unwanted results, called *electromagnetic interference (EMI)*. The technical field addressing these problems is called *electromagnetic*

---

* 1 hertz (Hz) is 1/60 s, or 16.7 ms.
† K. Min, *Collectives for Reducing Noise*, Paju, Korea: Sungandang, 1991, p. 347.
‡ C. Paul, *Introduction to Electromagnetic Compatibility*, New York: John Wiley & Sons, 2006, p. 840.

compatibility (EMC). More than 30 years ago, these issues were addressed in U.S. regulations by the Federal Communications Commission.* There are two considerations when dealing with EMI. The first, the susceptibility or immunity of the item, deals with its vulnerability to unexpected electromagnetic interference from other appliances. The second concerns electromagnetic emissions from the item itself influencing other appliances. There are various instances in which electromagnetic energy becomes interference, and therefore many kinds of regulations; there are also many facilities for measuring EMI, which requires a huge chamber. To check for EMI, it is best to seek the assistance of a specialized agency.

Another risk is the possibility of fire, with its attendant safety concerns. In testing for fire risk, a variety of abnormality tests should be executed. Because there is a danger of fire at a site where current flows at high voltage, everything from the power plug to the point of current extinction should be checked carefully. Regardless of how components are connected or disconnected, no spark or smoke should occur. Inserting a barrier block to isolate any fire should also be considered.

Safety concerns regarding electric shock can be checked by measuring leakage current, dielectric strength, and insulation resistance. Since dangerous current can flow through the human body when it touches an exposed metal part of an electrical appliance, the leakage current is regulated to less than a few hundred microamperes between metal parts and ground when the product is in operation. Insulation resistance should be greater than several megaohms when a few hundred volts are applied between the power line and any exposed metal. Dielectric strength is considered good if only a few milliamperes flow when several kilovolts are applied between the power line and exposed metal. Dielectric breakdown increases the leakage current and causes accidents.

Mechanically, spacecraft experience tremendous shock during launch and their structures undergo extreme thermal stresses during the shifts between outer space day and night, so these should be tested. Moreover, since the cosmic rays that are abundant in space create an electromagnetic environment that can interfere with electronic apparatus, especially in the area above the southern Atlantic Ocean near Brazil, causing abnormalities, space equipment should also be tested for this.

So far, we have discussed various approval tests, classified by conveyance and usage environments, and their bases. There are also tests that are specific to particular products and their environments. If colleagues working on a product meet and identify its various usage environments one by one, all the relevant test methods could be considered. To do this, they should thoroughly review the environmental and operational conditions under which the product operates and fully understand the

* Ibid., p. 11.

product's performance, structure, and materials. Based on this survey, they can then decide whether a certain test is required or not, and adjust the severity and sample size of the test. At this time the participation of a reliability specialist in identifying failure mechanisms will improve the process. None of the tests are unimportant, but their priority can be determined by failure rates and the amount of customer reimbursement related to failures.*

All the approval tests discussed so far are reliability marginal tests to identify trouble that occurs primarily in unusual or extraordinary environments. We have yet to look at tests confirming lifetime and failure rates under normal conditions. Next, we will discuss the methods and limits of frequently used tests and the reliability methods that can be substituted for actual tests. The reasons why reliability accidents occur in the products of advanced corporations will be described, as will the concepts of corrective activities and approval tests.

4.2   Reliability assurance methods and tests at the development stage, and ensuing problems

What reliability assurance methods are needed so that a product will pass these tests and work well for customers? Since the product is new, frequently used methods and tests will be discussed in sequence, according to product development stages.

If the pilot product is made according to concepts based on market demand, then tested, and if each of the problems exposed by the tests is fixed, the product can be made almost perfect, but it will be very expensive in terms of time and money. In order to reduce these costs, the stages of product development are sectioned into several levels, and various methods appropriate to each stage are used to review reliability and performance. Reviewing components or units at the earliest possible stage is preferable to testing a complete, assembled product. Many reliability-related methods have been developed that substitute for actual tests. Using these activities, called *proactive reliability engineering,* can dramatically reduce the expense of product development.

When such methods are introduced into the product development process, all reliability-related problems may seem to be solved, but these tests cannot completely eliminate all issues. Thus, they should be applied considering their limitations. Table 4.3 shows stages of product development and related methods and tests to ensure reliability for larger corporations. The underlining indicates actual tests.

---

* This can be estimated using Equation (5.1) in Chapter 5.

Table 4.3 Reliability Methods and Tests Matched with Stages of Product Development

Stage	CD DR#1	PD DR#2	DPP DR#3	LPP DR#4	TMP DR#5	MP
Concept	Product concept development	Product design	Desk pilot product	Line pilot product	Test mass product	Mass product
Proactive reliability engineering	• Customer/ market research • New feature survey • Extant issue survey	• Checklist • Search of failure cases • FMEA/FTA • FEA • Failure rate prediction • Derating	• Check heat source • Check moisture infiltration • Check cumulative contaminants • Check fatigue test • Check actual related tests	• Reliability marginal test • Stress-strength analysis • HALT	• Check MP-related change • Screening/aging test • Field test	

Note: The typical classification of development stages is shown in medium/large-sized manufacturing company and various reliability methods/ tests adequate to releveant stages are arranged.

First, let's consider what happens at each product development stage. The concept development (CD) for the target product is derived from the survey data of customers' needs and of the competitive products on the market and is then reviewed (DR#1: first design review). The product is then designed (PD) according to these concepts and reviewed (DR#2). Its prototype, called the *desk pilot product (DPP)*, is made and checked in terms of performance and reliability (DR#3). Then pilot products, called *line pilot products (LPPs)*, are made on production lines similar to the actual lines, with components made by real dies or molds, and reviewed for manufacturing problems and performance issues by various testing methods (DR#4). Finally, the test mass product (TMP) is produced on the real line and problems related to quantity production are investigated (DR#5). After these are cleared, mass production (MP) can begin.

The reason for dividing product development into stages for review is to reduce the failure costs of product development. Failure in DPP will create only product development expenses, but failure in LPP costs the investment in production tooling as well. Therefore, crossing from DPP to LPP is one of the chief executive's major decisions. One big enterprise in Japan appoints a chief technical executive in every factory, who checks technical issues in advance of moving to LPP from DPP and has a separate line of product handling authority from the strategic business division. He analyzes the results of DPP in his factory, lessening the risk of large investment and raising the likelihood of product success by requiring additional approval tests and so forth.

Some product developers want DPP components made by tools on the real line in order to expose potential issues there clearly, but such applications are unreasonably costly and should be avoided. DPP components are made using rapid prototyping and manufacturing (RP&M) technology, which can generate them without using regular tools like dies or molds. It is better to apply this method aggressively, because RP&M technology, used in making samples, and rapid tooling (RT) technology, a similar process applied to the production of small quantity lots, have been developing day by day.[*]

Now let's touch on reliability methods that accompany the various development stages. In the CD stage, new units are isolated and plans are made to check for the possibility of failure. At this stage, the product developer will also recheck for problems in the existing model or similar products.

During the PD stage, the design is thoroughly checked for omissions from the relevant design checklist, and its possible failures are reviewed according to the casebook compiling failures. The checklist should include

---

[*] A. Anderson, "An Automotive Perspective to Rapid Tooling," in *Rapid Tooling*, P. Hilton, ed., New York: Marcel Dekker, 2000, pp. 185–220.

the anticipated problems that might be revealed by the tests described in the middle of Section 4.4.

Failure modes and effects analysis (FMEA) and fault tree analysis (FTA) might be applied. These techniques are widely accepted, since FMEA has been effective in developing artificial satellites and FTA has contributed to the safety examination of missile launch systems in the United States.[*] FMEA helps the developer to select the most significant failure modes influencing the system and to take proper corrective actions. Its methodology, procedure, and written form are widespread. The important failure modes selected by FMEA can be handled by FTA. FTA produces a diagram of logic symbols with quantitative data showing the sequential chain of causes reaching to failure; the developer can exclude relevant failure modes in the design stage through FTA. In order to apply these two methods, the developer should carefully survey material-centered structures, one element of failure, and system performance, which both contribute to perfecting the quality of the product. But these methods, from failure mode prediction to corrective action, are not scientific, because the causal links are comparatively weak, resulting in quite different results than those that might be experienced by users. A specialist using reliability technology based on empirical experience related to failure mechanisms can increase the accuracy of these methods to improve the quality of the product, but this is insufficient unless it is backed up with scientific data.

For accurate failure prediction or secure corrective action, finite element analysis (FEA) can be simulated in a related structure, a method that has long been used for solving performance problems. In order to reproduce the same phenomenon that will occur in actual use and increase the accuracy of the analysis, the material characteristics and boundary conditions should be determined with real data. Simulation packages proven by real fields produce more accurate predictions because the input data and conditions in simulation results deviate from those established by real phenomena. Since the performance of computer and software related to FEA has advanced greatly of late, the increased numbers and applications of field-proven simulation packages will contribute significantly to developing the next model. FEA helps the developer search for weaker sites and their marginal limits—these are, however, not the same as lifetime and failure rate. Linking simulation results with reliability indices is a challenge remaining to be solved in the future.

In the case of printed circuit assemblies, the failure rate will be predicted by analyzing stresses from outside. An exclusive software package determines the failure rate or the mean time to failure (MTTF) by

---

[*] D. Park, J. Baek, et al., *Reliability Engineering*, Seoul, Korea: Korea National Open University Press, 2005, pp. 196, 203.

considering stresses due to environmental and operational conditions. (But again, note that the MTTF is not the same as the lifetime, and its inverse, or failure rate, is meaningful.) This software is designed based on MIL-HDBK-217 (the *Military Handbook for Reliability Prediction of Electronic Equipment*), but its prediction results do not completely reflect the real situation,[*] which raises questions about its basic postulates. Consequently, the U.S. Army announced it would abandon its use and instead apply physics of failure methods; nonetheless, other organizations of the Department of Defense still use it because of the benefits of quantification.[†] This method should therefore be revised to match the real circumstances of each company before it can be usefully applied. For mechanical units, it is impossible to systemize such a method because minor changes in material and structure, such as defect removal or corner radius adjustment, can produce large differences in the failure rate.

In the case of electronic components, derating standards are applied to circuit design. Derating means that the designer decreases the rated value written into the part specifications for suppliers, which equates to the inverse of the safety factor in mechanical systems. For example, 50% derating equals a safety factor of 2.[‡] As you might guess, derating parameters differ from one component to another, and derating rates also vary according to the use environment. For example, the derating parameters of a connector are voltage and current; current will be derated to 85% for office use, or to 70% for a vehicle.[§] Derating will decrease the failure rate but increase cost, since designers should select upgraded components to ensure the desired safety margin.

At the DPP stage, when a real product exists, the developer investigates any spot where temperature is excessively elevated. If there is a local hot spot in the printed circuit assembly (PCA), neighboring devices will degrade rapidly. Therefore, it is good to take a picture by infrared camera to survey temperature distribution throughout the PCA while it is operating. The junction temperature of electronic devices such as integrated circuits (ICs) should usually be below 125°C, above which the failure rate increases rapidly. Decreasing by several tens of degrees from 125°C will slow degradation.[¶] The developer must pay attention to temperature rises in circuit devices, since this is a significant factor affecting the reliability of the PCA. In response, he should design the structure to exhaust heat from the heat sink by natural convection. Otherwise, forced convection

[*] See Chapter 2, Section 2.6.

[†] A. Amerasekera and N. Farid, *Failure Mechanisms in Semiconductor Devices*, New York: John Wiley & Sons, 1998, p. 272.

[‡] B. Dudley, *Electronic Derating for Optimum Performance*, Rome, NY: Reliability Analysis Center, 2000, p. 2.

[§] Ibid., p. 58.

[¶] Ibid., p. 171.

must be used, which will increase the cost. Various standards limit the temperature of electronic devices under use conditions.

The product developer also needs to check the infiltration of moisture and survey its effects. Moisture ionizes the material, turning it into another compound, while ambient heat raises its internal energy, accelerating the transformation and weakening the material. To confirm this phenomenon requires introducing ample moisture during testing, because excessive heat dries up the moisture and prevents observation of its effects.

The developer must also assess the accumulation of dust and contamination over time. Again, moisture particularly infiltrates the porosity of the dust layer and creates electrical shorts, which can cause fire accidents.

The developer should also confirm whether the actual tests of core units, including components related to fatigue, were executed by the vendor, and survey the validity of the test methods, conditions, duration, sample size, and results.

If there is no problem in the DPP stage, the developer reports to the CEO that product design is completed. Thereafter, problems related to productivity will be assessed and large investments can be made in such items as jigs, fixtures, dies, and molds.

At the LPP stage, the various reliability marginal tests discussed in the previous section can be performed, since there are now many available samples. Stress-strength analysis should be conducted at weak sites, measuring material strengths and stresses on the structure.[*] With this, the failure rates due to overstress failure can be quantitatively estimated, keeping in mind that failure rates apply to the initial material strength before any degradation. Of course, if the material strength is checked after a certain period of operation, the acquired data represent the degradation during that period, but they cannot be applied to the potential additional material change after the observed period.

Through highly accelerated life tests (HALTs), weak areas in the item can be identified.[†] The basic concept of this test is to speed up thermal shock and vibration, which can easily lead to broken links—for example, fatigue ruptures of PCA solder joints.[‡] This is a useful method for rapidly searching for poor design and defective materials that become evident under severe operating conditions. But with HALT, there is as a great possibility of precipitating atypical failures as there is of finding those likely to occur under expected operating conditions—a factor that should be considered when taking corrective action. Another point to remember is

---

[*] P. O'Connor, *Practical Reliability Engineering*, New York: John Wiley & Sons, 2002, p. 114.
[†] G. Hobbs, *Accelerated Reliability Engineering*, New York: John Wiley & Sons, 2000, p. 5.
[‡] Thermal shock results when an item endures repetitive cycling from high temperature to low temperature and vice versa. Thermal shock degrades solder materials and breaks them due to repetitive expansion and compression, called *solder joint fatigue*.

that the accelerated aspect of HALT is not intended to identify all failure modes. For example, minute transformations due to chemical reactions in the presence of moisture cannot be identified through this test because such reactions are activated more by high temperature than by thermal shock. There is also the possibility that transformed material will disappear or that chemical reactions will be retarded due to harsh vibrations. In addition, HALT equipment uses liquid nitrogen rather than refrigeration when freezing test items, which abruptly lowers the ambient temperature and saves test time but also costs more.

In the TMP stage, reliability issues occurring in production are checked as production quantity increases to the planned volume. Additionally, as working conditions—such as production facilities, molds, or operators—change, the possibility of item failure must continue to be checked. Speeding the cycle time of the manufacturing process will lower the material strength and distort items molded by injection machines, as well as those in the assembly line that experience insufficient or excessive fastening strength.

Defective items can also be identified through aging and screening tests, and are generally released to the next processing stage if they prove good. However, although the distribution of item failure may be assessed if many items are tested, the results will reflect infant mortality, not failure during longer use. If test conditions are severe to normal, real data about operation in use can be obtained, but test conditions themselves also produce aging or damage in the item. Therefore, while the causes of problems identified by these tests must be clarified and corrected, such tests should not be performed in the mass production stage. Note that the tests should continue if they identify problems and mass production should be delayed.

Another approach is to observe the customer use results from a confined test market over a certain period of time. Again, with this method item problems under various environmental and operational conditions may be identified, but they would pertain to the infant mortality of the item, not wear-out failure under normal operating conditions. It is not difficult to discern the constituent elements of a field test, like sample size, observation period, and limits.* The field test ought to identify the types of failure as well as the failure rates. Using these results, the manager will determine the period over which the field test should be conducted. Once the test is concluded, approval tests of the product should be reviewed again, including accelerated life tests if any major wear-out failure is found. This is because the field test is generally performed under normal (not severe) conditions, and thus not all problems can be found without waiting a lengthy period.

---

* See Chapter 5, Section 5.7.

Since the proactive reliability methods discussed above are not easy to apply, the recommendation is to mix and match methods appropriate to the product and to the company's competence. Making them as simple as possible and performing them early in the product development process will reduce the lead time for product development. At this point, consulting with a reliability specialist experienced in failure mechanisms can be very effective.

The more components there are in the product, the greater will be the material costs and expenses. Likewise, there will be more problems as the number of parts grows.* Thus, a good developer always strives to minimize the number of needed components, considering in detail the function of each item. Of course, the designer also considers adopting cheaper materials and reducing material weight, but these steps may be detrimental to reliability.

Finally, most companies seek to graft the reliability methods of advanced companies onto their own processes; however, sometimes they place too much confidence in them. If the bases of reliability and the limitations of these methods are not understood and considered, many methods will seem rational although they are not scientific, as explained in the next section. Even if the same equipment used by advanced companies is introduced, difficulties will arise when establishing specifications related to reliability approval. In the end, companies tend to take the easy way, which means that they adopt the same units or components used by successful companies. Superficially, this seems to be a good method, but it increases costs and procurement is sometimes delayed. It is not a first-class strategy to merely adopt the same methods, equipment, and components of the most advanced producers, rather than developing methods specific to the company's needs.

4.3   Why reliability accidents occur in the products of advanced corporations

So far, we have reviewed various reliability methods and tests and their limitations. Let's look at a case study. In 1966, the new product development process used for the Boeing 777, a symbol of U.S. technological excellence, was reported by Swink.† Signifying the importance of teamwork and communication among parallel functional organizations, the first 777 produced was named "Working Together." Communication was a priority not only between internal groups, but also with customers and suppliers. Representatives from the first four major airlines to

---

* See Chapter 3, Section 3.3.
† M. Swink, "Customizing Concurrent Engineering Processes: Five Case Studies," *Journal of Product Innovation Management*, 13, 1996, 234.

purchase the 777 participated as members of the development team. In addition, Boeing's "design-build" teams included representatives from many of the project's approximately 100 major suppliers. Boeing intensively used three-dimensional digital design technology to simulate the performances of its units and confirm stresses at weak sites corresponding to possible failures. For these analyses, 1,700 computer workstations in Seattle, more than 500 elsewhere in the United States, and 220 in Japan were mobilized. Furthermore, to prevent failures, the 777 program utilized only field-proven technologies, backed by extensive testing. Physical prototypes, after execution performance and stress analyses using the three-dimensional design system, were laboratory tested under severe environmental conditions before any parts or systems were incorporated into the first aircraft. With the test results, designers addressed imperfections in the design and maximized its reliability and durability. This heavy emphasis on testing was the basis for the success of the 777.

Why are such units rigorously tested even after reliability methods are applied? The answer is that problems occurring in units in real situations cannot be fully anticipated even though analyses of reliability methods yield good conclusions, because there are so many parameters or unconsidered variables. If there were no specialists conforming the relationship between methods and real conditions, or no other data clarifying these relationships, the results of applying reliability techniques would be useless.* Moreover, if innovative structures or new materials are adopted, even experienced specialists cannot figure out what possible new failure may lie dormant in them. Especially for electronic items, it is already difficult to isolate problems because of the minute structures and multiple materials used. Electronic items include such diverse materials as silicon crystal, gold wire, plastics, rubbers, and adhesives, while mechanical components generally comprise only metals, lubricants, and paints. Although we can attempt to analyze electronic items with finite element analysis (FEA), the many characteristics of electronic materials are not fully known. Even if all these characteristics were identified and researched, the amount of work to perform FEA on the multiple tiny structures is prohibitive.

Thus, tests of the item in actual use are necessary. To improve the production yield rate of semiconductors, statistical experiments are considered more effective than the technology intrinsic to microcircuit devices; this statistical methodology is called the *Taguchi method.*† In mechanical

---

* Y. Ahn, "Strategic Collaboration Analysis and Policy Issues for Co-development of Aircraft T-50 between Korea and U.S.," Seoul, Korea: Korea Institute for Industrial Economics & Trade, 2007, p. 75.

† S. Taguchi, "The State of Education and Dissemination of Taguchi Method in U.S.," in *Robust Design Case Study; U.S. and Europe* (Robust Design Seminar 6) (Korean edition), Seoul, Korea: Korean Standards Association, 1991, p. 371.

units, durability can be reduced by the degradation of nonmetal parts, such as rubber, plastic, and grease, even when the metallic components in the units hold up well. The fatigue life of mechanical components differs significantly due to surface roughness. The reliability of these items cannot be properly assessed using only reliability-related methods, but can be estimated using the results of actual tests on physical models.

Again, the Boeing 777 program utilized only field-proven technologies, backed by extensive testing, to maximize durability and reliability. We cannot know how efficient these tests were or how much waste was incurred, but we do know that physical tests are the right way to maximize reliability. Although advanced corporations adopt and perform various methods and tests, their products still experience failures that are widely reported in the media. Let's take a look at some known reliability failures.

As mentioned in Chapter 1, rotary compressor accidents occurred in the refrigerators of U.S. company G and company M of Japan in the 1980s. Moreover, there was the explosion of the space shuttle *Challenger* in 1986. In another case, Company D-C came near to bankruptcy due to lawsuits over the failures of its breast augmentation product. Company F in the United States experienced difficulties with rollover accidents, especially in four-wheel-drive vehicles, due to problems with the tires.[*] In 2000, the reputation of leading company S in Japan was tarnished due to failures of charge-coupled devices incorporated into over 10 million cameras it produced; company M in Japan lost credibility due to the leakage of carbon monoxide in the rubber hose of a forced-flue stove, which caused the death of some users.[†] Recently there were battery accidents in the products of company S in Japan, recalls due to defects in the telematics of leading automaker B in Germany, and a recall of popular models from automaker H in Japan.[‡] Automaker T in Japan, which had a good reputation for durability, had to focus on reinforcing its quality organization after issuing massive recalls. Quality accidents related to reliability, which occur despite the glorious triumphs in science and technology that have put humans on the moon, send us back to the fundamentals to consider what product quality truly is.

Why do reliability accidents occur in products designed and manufactured by advanced corporations? Why can't sophisticated corporations eliminate these large (and embarrassing) failures? Physical test methods are established through the review of failure case studies, and some of them are now prevalent in industry. Are there mistakes in applying these test methods? Or is there a need to complement the tests? Let's consider

---

[*] K. Kim, "Lesson from the Recall of Sony Battery," *LG Business Insight*, November 22, 2006, p. 17.

[†] See Chapter 7, Section 7.3, footnote * (p. 146), and Chapter 5, Section 5.2, footnote * (p. 92).

[‡] K. Kam, "Establish the Basics of Management," *LG Business Insight*, February 2, 2007, p. 17.

scrutinizing the way we *think* about the test methods, including the basic concepts that underlie them, rather than investigating individual causes of failure.

Under the suggested approach, engineering would be tasked to design products on the bases of relevant theories and data, with as few confirming tests as possible. After all, since the design data have been taken from the results of tests and experiments related to the products, the occurrence of failures connotes imperfections in the data, which means that some areas related to the tests were misunderstood.

There are two underlying problems with the test-oriented approach.[*] First, we cannot identify reliability problems inherent in new structures and materials using existing test specifications. The task of product engineers is to improve performance and lower the cost of products. To do this, the engineer must adopt new materials and components and devise innovative and creative structures. But doing this alters inherent failure mechanisms or introduces new ones as the two elements of failure mechanics, stress and material, are changed. Altered failure mechanisms can hardly be found by the test methods applied to products before the alterations. Therefore, it is necessary to review the reliability test specifications, or perhaps to establish new ones, because the existing specifications cannot be matched to new items. We should also consider what new tests might be useful. There are no absolute and permanent test specifications for reliability. The changed relationship between stress and material requires the alteration of test specifications.

Second, whether or not the products meet the necessary quantitative reliability targets, lifetime and annual failure rate within lifetime cannot be identified, since existing tests for reliability specify test conditions, sample sizes, and permissible numbers of failures, but then express the results as simply pass or fail. Efforts to calculate the lifetime and annual failure rate within the lifetime necessarily fail because the bases for establishing the test specifications are not clearly explained or understood. If a problem with an item is detected by current test methods and corrective remedies are adopted until the improved item passes the same tests, we still do not know whether the reliability of the item meets the identified quantitative targets. At the very least, the reliability index under normal usage conditions for most customers should be confirmed quantitatively. In other words, the lifetime of the weakest site on the item should be confirmed as no less than the lifetime target of the system. To do this effectively, the bases of test methods must be reconfigured to provide quantitative results that can be calculated and checked against numerical targets. If these targets are not met, the item should be corrected and tested again. Thus, test methods need to be changed so that quantitative conclusions

[*] See Chapter 1, Section 1.3.

can be drawn to ensure that the item will work until the lifetime expected by customers. The conclusions of reliability test methods should express, for example, the prevailing environmental conditions, the identified failure site(s), the operative failure mechanism(s), and the lifetime until the determined cumulative failure rate is reached (BX life, or below X percent of cumulative failure rate for Y years). Note that such quantitative verifications are valid under normal environmental and operational conditions, but not under unusual conditions.

Predictive power diminishes when there are a moderately large number of factors and interactions in a certain phenomenon.[*] But if we separate the reliability of the item into lifetime and failure rate within lifetime, the factors related to reliability estimation are decreased and statistical quantitative verifications of the item can be obtained.[†]

Consider again the rotary compressor failure in refrigerators.[‡] Let's enlarge our understanding of the structures that change in accordance with the two elements of failure mechanics, stress and material. The rotary-type compressor has a longer gap between sliding metal components of various materials, along which high-pressure gas leaks can occur, than does the reciprocating type. This gap makes it difficult to produce high compression. This is why rotary compressors are used for air conditioning systems, which require only cool air, and reciprocating compressors, with their shorter leakage lines and greater compression, are incorporated into refrigerators, which need freezing air. Of course, rotary compressors can produce very cold air, but the technology to do that is highly exacting, which means that there are more critical-to-quality factors (CTQs) than for a reciprocating compressor. Note that, in this case, it is crucial to find and manage all the attributes influencing leakage, hampering lubrication, and causing wear due to the higher compression ratio and longer leakage line.

Company G established a life test specification that runs continuously for 2 months and supposedly simulates 5 years' actual use.[§] With this specification, 600 compressors were tested and not a single failure was detected. Consequently, over a million new compressors were released by the end of 1986. Within a year an avalanche of failure was reported. Although the materials had been changed from cast iron and

---

[*] P. O'Connor, *Practical Reliability Engineering*, New York: John Wiley & Sons, 2002, p. 159.
[†] See Chapter 5.
[‡] See Chapter 1, Section 1.3, and above.
[§] T. Bahill and S. Henderson, *Requirements Development, Verification, and Validation Exhibited in Famous Failures*, New York: Wiley InterScience, 2004, p. 6. This is a dangerous way of thinking as far as wear-out failure is concerned. When a large quantity of an item has been sold, if epidemic failure occurs in 5 years, the expense of handling it will be prohibitive. Test specifications for refrigerators should be matched to anticipated usage periods, or over 10 years.

steel to powdered metal—a completely different structure—reliability engineers thought the existing specifications would still be appropriate to test compressors incorporating these new, cost-saving materials. The large number of test samples, perhaps, made them comfortable about reliability, but the resulting failures within a year indicate that the existing test specifications, anticipated to prove at least 5 years of operating life, were completely inapplicable for the new compressors. This highlights the importance of confirming the validity of reliability test specifications, as described above in the first condition.

Company M's refrigerators experienced epidemic failures over several years after the inclusion of new rotary compressors that incorporated design changes for cost reductions.[*] Reliability engineers should have realized that minor design changes and even trivial deviations in the manufacturing process could greatly influence the reliability of rotary compressors, which have a relatively weaker structure than reciprocating ones. They could have identified the problems in advance if life tests were conducted on whether the B1 life exceeded 10 years and if failure rate tests were performed with an adequate sample size for the target. This points to the importance of quantitative confirmation of reliability test specifications, as described in the second condition.

Let's consider the outline of quantitative test specifications, using common sense and mental arithmetic (the basis of which will be explained in Chapter 5). If the lifetime target is B1 life 10 years, it is satisfactory if there are no failures in a test of 100 compressors (that is, the cumulative failure rate is 1%) tested for 4 months (120 months/30) under severe conditions leveled up by an acceleration factor of 30.[†] Note that the sample size can be decreased to approximately 25 if the test period is extended to two times the period, or 8 months.[‡] For a failure rate target of 0.3% for 3 years for all CTQs, 330 compressors would be tested for 2 months (36 months/20) under severe conditions, leveled up by an acceleration factor of 20, with the condition that no failures are found. If there is concern about overstress failure due to material degradation, the above test would be changed, for a failure rate target of 0.5% per 5 years, to 200 compressors tested for 3 months (60 months/20) under severe conditions, leveled up by an acceleration factor of 20, again with the condition of no failures being found. In the case of failure rate confirmation, random sampling can help uncover various problems corresponding to material

[*] Communications in Reliability Seminar arranged by Matsushita Consulting at Kadoma, Japan in 1999.
[†] Since the failure rate, after testing, can be estimated with a commonsense level of confidence, (failed number + 1)/(sample size) (see Chapter 2, Section 2.3), it is sufficient to test 100 with the condition of nonfailure found for a failure rate of 1%, 330 for a 0.3% failure rate, and 200 for a 0.5% failure rate.
[‡] See Chapter 5, Section 5.6.

defects and process deficiencies. Now let's figure out the test conditions. It approximates actual usage to assume that compressors are operating intermittently, which causes them to work in a state of insufficient lubrication, with increased pressure differences under higher temperatures.

If CEOs grasped this concept and could mentally interpret the test reports, they could understand the framework of test specifications and assess their validity. If they assume their engineers are finding and fixing reliability failures, they feel comfortable that the test specifications are effective; if there is no failure found, they can request complements to the test specifications, such as retesting, extending the test period, or leveling up the severity of the test conditions.

Quality accidents related to reliability will disappear if the two original problems—applying adequate specifications to new products and quantifying test results to the reliability target, instead of simply rating them pass or fail—are settled. New test specifications that arrive at quantitative conclusions with clarified bases should be designed according to the newly designed structure. Note that understanding an item's structure, the environmental and operational conditions under which it works, the degradation of its constituent materials, and statistical knowledge are all important for establishing revised test specifications in quantitative form.

The number of test specifications for reliability tests has increased only gradually, since they are difficult to abolish or revise due to constantly changing applications; the limits of the specifications are not recognized because their foundations are obscured by pass/fail evaluations. In order to set adequate new test specifications when the structure of an item has been altered, test specification planners always create their own methods, because there are few standards to refer to, which is another issue. Moreover, there are few experts to review them and no explanation of foundational principles is established for the existing specifications. While the development of products is proceeding rapidly, the progress of design methodology for test specifications is virtually at a standstill.

Becoming an advanced corporation requires developing original technology, which means adapting creative structures or new materials. These new technologies increase the probability of reliability-related failures. Because of the lack of existing test methods for reference, the development of verification methods is an absolute prerequisite to the development of products with original technology.

The product development chief will be congratulated if his product prevails in the marketplace due to improved performance or cost reductions. In order to succeed rapidly in the market with the introduction of original technology and to reach the level of an advanced corporation, rapid product verification must be perfected. Thus, developers of test methods that can substantially reduce the time to market should also be rewarded, because their newly developed test methods can be applied to consecutive models to achieve market domination.

4.4 New classification of reliability assurance tests

The task of establishing new quantitative test specifications adequate for innovative items seems quite daunting. But not all the old specifications need to be changed. After classifying and synthesizing all the test specifications available as of now, let's reconsider and quantify them.

Although the processes described in Section 4.2 are carried out at each development stage, product reliability should finally be confirmed with actual tests if the methods used at each development stage are not proven technologies. As explained in Chapter 1, Section 1.3, we aggregate existing issues, anticipated issues, and comparative disadvantages and then divide them into two groups: performance issues and reliability issues. Sometimes it is difficult to divide issues into these two categories. The criterion for this division is whether or not any physical change occurs in the materials and ensuing structures.

Performance issues should be further divided into common performance issues and special (environmental) performance issues; doing this enables approval engineers to consider all necessary tests. Special performance issues occur under unusual environments, such as elevated electromagnetic fields. The necessity of a certain performance test can be checked easily if the environmental conditions under which the items function are investigated in detail.

Reliability issues can also be divided into two groups: reliability marginal issues (unusual environments) and reliability quantitative issues (lifetime and annual failure rate within lifetime). Reliability marginal issues include physical changes in items under severe or peculiar conditions, including unusual usage environments, such as electrostatic overstress, lightning surge, and storage and transportation.* Reliability quantitative issues include reviewing and estimating the lifetime and annual failure rate within lifetime under almost normal conditions, which is the measure of the second problem discussed in Section 4.3. Most existing test specifications can be classified as reliability marginal tests, and reliability quantitative tests can be set up as new tasks in a test plan of BX lifetime.† Figure 4.1 illustrates this classification.

Some readers who understand the complexity of reliability tests wonder at the simplicity of the categorization in Figure 4.1. Reliability test methods have been developed into various types in, for example, Japan

---

* See Chapter 4, Section 4.1.
† See Chapter 5, Section 5.2.

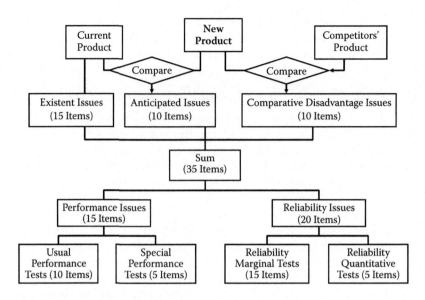

Figure 4.1 Conceptual framework for quality assurance.

Standard JIS Z 8115.* There are a great many kinds of test methods since each test responds to one failure accident that has occurred in the past. It is not easy to understand test concepts and remember their terminology, because so many tests are similar. Actually, two types of methods, marginal and quantitative ones, sufficiently comprise reliability tests, as shown in Figure 4.1.

Issues with new products, whether already existing, anticipated, or matters of comparative disadvantage, should be investigated in depth. What kinds of stresses produced from conveyance or use environments and operational conditions would be imposed on the product structure or materials, from the introduction of the innovative design until the product is scrapped? How are the stresses dispersed and absorbed in the new structure? With the results of this careful analysis, necessary new tests can be identified, along with appropriate condition levels to fully test the product. Doing this in detail ensures that the selected tests are confirmed and all conveyance and use environments and operational conditions are accounted for. It also means that all the likely stresses experienced by the structure and material of the item are considered, anticipating potential problems with the unit structure of the item.

The next step is to review what kinds of proactive reliability engineering (as discussed in Section 4.2) can be applied to ameliorate possible

* H. Shiomi, "Reliability Test," in *The Reliability Handbook: Technology toward Twenty First Century* (Reliability Engineering Association of Japan), Tokyo, Japan: Union of Japanese Scientists and Engineers, 1997, p. 194.

troubles before the actual tests. The CEO should ask the chief of product approval to study the relationships between the anticipated issues and proactive reliability methods and should require the chief of product development to confirm the results of the methods used according to this study.

Now let's consider actual tests for new product approval. As usual with tests that check performance under normal conditions, the sample size can be small unless it is necessary to confirm the average value and the standard deviation of performance characteristics distribution. With special performance tests and reliability marginal tests, it is enough to sample a few items or test for a relatively short period. When the probability of peculiar environments is low and the affected number of products in such instances is low (with the cost of reimbursements for failed products correspondingly low), the typical pass/fail test is sufficient.

But one thing needs to be pointed out. If no trouble is detected when testing four sets in a certain special performance test, it means that the quality defect rate reaches a maximum of 25% with a commonsense level of confidence. Moreover, when no failure occurs in testing five sets for 72 h in a certain reliability marginal test, the failure rate per 72 h is guaranteed to be a maximum of 20% with a commonsense level of confidence. Therefore, it is unreasonable to approve testing items with small samples if there is a significant chance that they will experience stresses from special environments.*

Now let's consider lifetime testing via reliability quantitative tests under normal conditions. If the material undergoes large stresses, it will break; if it receives small stresses repetitively, it will also break in the end. Breakage due to large stresses can be detected without difficulty. Failure due to the accumulation of small stresses cannot be easily detected ahead of time, so lifetime tests are required. As noted, business survival would be impossible if products demonstrated epidemic failures in 2 or 3 years under normal environmental and operational conditions—especially for durables, which are expected to provide 10 years of failure-free use. Thus, every unit—the aggregate components—of a finished product should be tested in order to quantitatively estimate the lifetime in terms of possible failure modes.†

Testing many samples or doing so for a long time is not the same as a lifetime test. Careful preparation is necessary in order to calculate an item's lifetime. Quantitative information results from scientific preparations for designing test specification, such as establishing severe conditions

---

* The basis and mental arithmetic for this conclusion will be explained in Chapter 5, Sections 5.6 and 5.7.
† Setting up a test plan for units of complex products will be discussed in Chapter 5, Section 5.2.

responding to actual stresses, computing the test duration to assess material degradation, and calculating a sample size adequate to determine the BX lifetime target. Furthermore, testing should be designed to produce failure with the same failure mechanism that affects the actual product, reviewing in detail the parameters related to the applicable stresses. The results should be expressed as the BX life Y years, which means that the period to reach the cumulative failure rate of X percent is Y years.

Three categories of test conditions are recommended for lifetime testing. In the first type, the test runs continuously, searching for material degradations or wear-out. The second type simulates nonoperation, leaving items alone in typical environments that induce chemical changes, such as corrosion. The third type induces repetitive working and stopping conditions to check for disorders due to an imbalance of power dissipation. The failures tested under these conditions can be analogized with those of the human body. The first is like testing a runner who quits midway during a marathon, the second tests for a person's inability to function normally after a lengthy stay in a hospital bed, and the third parallels sitting down on a soccer field after several short sprints. Thus, we need to find out, quantitatively, in what time a certain runner can finish the full marathon course, how soon he can return to work under normal conditions after extended bed rest, and how many sprints he can run in 90 min. For example, we can estimate the distance and time at which the runner quit the marathon or the time to complete the race, with only 20 min of testing, for example. We can reach that goal if the items are tested according to new test specifications established considering the above three conditions.

Let's consider testing for the failure rate within lifetime—that is, reliability quantitative tests. To establish this exactly, the two elements of stress and materials must be examined thoroughly. For materials, this includes reviewing what kinds of defects or mistakes might be built into the structures, and surveying manufacturing procedures from raw materials to shaped materials as constituents of structure.* For stresses, it also includes reviewing what kinds of normal and abnormal stresses would be imposed on items throughout their lifetime, then deciding whether such stresses would be absorbed or negligible. Materials in the product that absorb stresses must be evaluated as to whether they can withstand them. If the conditions or stresses that produce failure are peculiar, the product must be tested under different conditions than those used for normal lifetime tests. For example, in the case of materials partially weakened over about 70% of the lifetime, it is important whether a particular unusual stress is absorbed or not. If it is likely to be absorbed, the failure rate tests are designed like lifetime tests, including material degradation.

* See Chapter 5, Section 5.7.

So far, we have discussed the inadequacy of existing reliability test specifications due to misunderstood reliability concepts and the need to set up new concepts to underlie reliability quantitative tests. The fact that the core of design technology is the design of approval test specifications, as explained earlier, also emphasizes the importance of reliability quantitative tests. Such testing is the answer to the proposition that test specifications should be reestablished, with secure bases adequate to newly designed structures, in order to compute quantitative results. We call this a *parametric accelerated life test (parametric ALT)*, referring to the need to first establish usage parameters. The core concepts of parametric ALT, the fundamental solution for reliability approval, are discussed in the next chapter.

chapter five

The fundamental solution for reliability assurance:
Parametric ALT

Our task is to reestablish reliability quantitative test specifications after conducting reliability marginal tests. As we said before, reliability test specifications should be designed concretely in order to establish quantitative definitions of the lifetime and annual failure rate within lifetime. Quantitative indices of item reliability should meet the targets established with reference to the quality of competitive items that customers recognize in the market. Parametric accelerated life testing (ALT) is the methodology for verifying whether the item's reliability meets those targets.

There are two kinds of targets: absolute targets and relative targets. A relative target is changeable relative to corporate circumstances for an ongoing concern, such as annual revenue or the scale of the company. The absolute target cannot be varied, however, and should be adhered to because of absolutes in the market, like selling price or after-sales service rate. If, for example, the selling price of a certain item is higher than that of its competitors' products, the item will not successfully penetrate the market. Similarly, if the after-sales service rate is higher than that of competitors, the item will not generate a profit. Sometimes an epidemic failure will bankrupt a company due to massive recalls and scrapping. Thus, the costs of materials and manufacture must be planned with respect to the adequacy of the selling price. The quality plan, especially the reliability plan, must produce products that equal or exceed those of advanced companies, and should also be established quantitatively. Finally, the budget for after-sales expenses should be carefully allocated according to the test results.

We will discuss how to establish a reliability plan for an item and to design parametric ALT. Setting the reliability target for the item and the two core concepts of parametric ALT—test period decrease and sample size reduction—will be considered in turn.

5.1 Benefits of quantified tests and parametric ALT

Parametric ALT uses quantified test specifications to extract quantitative results, and it can be easily validated, confirming its adequacy with specified parameters.

Quantification should be the basic datum of all judgments. The pros and cons about new product development can be debated endlessly between the design section and the assurance section with nonquantitative data. Assurance personnel want to ensure that the product will not fail due to the failure mode revealed in the in-house test. Design people reply that corrective action is unnecessary for a failure that occurs under severe, abnormal conditions, because such conditions are not likely in real use. It is also difficult for CEOs to blame the designers for such failures, given that they struggle to meet scheduled, often short, timeframes under pressure to meet the assigned cost limit. If CEOs were aware of these designer/assurance section arguments, they would be forced to make policy decisions to resolve the issue. As discord between the two sections increases, CEOs sometimes order the section chiefs to change positions with each other to promote mutual understanding or to commit the task of product approval to the product development chief. Or perhaps the CEO moves product approval to a higher authority than the section chief, hoping for staff harmony and a trouble-free product. But when one group is favored by the CEO's policy decision, the morale of the people recommending the other approach is compromised. These are all band-aid solutions, and real problems cannot be solved with them. The product is put at risk when organizational harmony is the prime decision factor. Such decisions are rational, but most of them are not made on scientific bases. In order to avoid these problems, the CEO should try to make such decisions using objective information. These issues can only be solved by using quantitative data developed by building reliability technology capability.

Some cases can serve as examples. Let's say testing established the failure mode and quantitative results for a unit with 100 components incorporated into a certain home appliance that has a total of 1,000 components. The appliance is expected to be used for 10 years. Let's decide whether a design change is necessary or not, remembering that the accumulated failure rate target for 10 years is 1%, as shown in Table 3.5. Consider the following scenarios:

1. This failure mode occurs after 13 years.
2. This failure mode occurs annually at a rate of 0.01% after 3 years.
3. This failure mode occurs annually at a rate of 1% after 3 years.

4. This failure mode occurs annually at a rate of 0.01% after 3 years, and ensuing accidents due to failures are likely to result in product liability (PL) issues, including injuries to users or property, for which the company is held responsible.

We can easily obtain appropriate answers to whether a design change is needed in the above situations.* Moreover, since future situations could be accounted for with the quantified data, design changes could be made by lower-level staff and section chiefs would need only to confirm the bases of the data, so mutual cooperation would flourish.

Let's consider other benefits of quantification. When we know whether the reliability of an item exceeds the given target or not, a decision about product release into the market can be made easily. Moreover, the effects of improving the product can be calculated exactly knowing the lifetime and the annual failure rate within lifetime. If the targets are well exceeded, perhaps the cost of the unit can be planned for its reduction.

It is important to calculate the warranty costs (C_W) after product release, which can be done in advance. The equation for this is as follows:†

$$C_W = (\lambda_1 \cdot E_1 + \lambda_2 \cdot E_2 + \cdots) \cdot Q \cdot L = (x_1 \cdot E_1 + x_2 \cdot E_2 + \cdots) \cdot Q \qquad (5.1)$$

where λ_1, λ_2 are average annual failure rates corresponding to the failure mechanisms, E_1, E_2 are repair and reimbursement costs related to each failure, Q is the annual sales quantity, L is the warranty period or lifetime, and x_1, x_2 are the cumulative failure rates to the end of the warranty period or lifetime ($x = \lambda \cdot L$). Dividing the warranty costs (C_W) by the annual sales amounts ($Q \cdot P$) yields the warranty cost ratio to sales amount (a), as follows:

$$a = \lambda_1 \cdot L \cdot \frac{E_1}{P} + \lambda_2 \cdot L \cdot \frac{E_2}{P} + \cdots = x_1 \cdot \frac{E_1}{P} + x_2 \cdot \frac{E_2}{P} + \cdots \qquad (5.2)$$

where P is the unit price of the product. If there is no trouble after one use because there is no accumulation effect, as for a disposable commodity, the above equation is changed as follows:

---

* In case 1 there need not be a design change since failure occurs after 10 years of use (expected lifetime). In case 2, there is also no need for change because the cumulative failure rate from 7 years to lifetime reaches only 0.07%. In case 3 a design change is needed because the cumulative failure rate from 7 years to lifetime reaches 7%. And in case 4 a design change is needed because the flaw is a critical issue.

† Assume that annual sales quantity is constant.

$$a = f_1 \cdot \frac{E_1}{P} + f_2 \cdot \frac{E_2}{P} + \cdots \qquad (5.3)$$

where f_1, f_2 are the failure rates of the disposable item.

5.2 Target setting for parametric ALT

Buyers demand compensation if the cumulative failure rate within the warranty period exceeds a certain level. Although buyers naturally request the maximum rate in order to keep their customers happy, the manufacturer should provide a fair warranty and continue to try to lower the rate in order to increase the customer base. Many companies guarantee the failure rate for 1 year, regardless of the warranty period. If there are no epidemic failures after 1 year or the warranty period, the company feels relieved.[*] A product that fails after the warranty period is repaired at the user's expense, but if there are a lot of such cases, word of mouth depresses demand for that brand and the market share decreases. Companies that want to become world class go the extra mile to ensure the reliability of items beyond the warranty period up to the lifetime customers expect. This can be accomplished if the overall plan of parametric ALT discussed below is established and if test specifications are designed in accordance with the concepts of parametric ALT and executed properly.

Generally, reliability tests proceed as follows. First, reliability marginal tests, the sample size for which is low, are carried out. Then an overall plan of parametric ALT for checking the lifetime of units is established. Finally, apart from the lifetime tests, parametric ALT is implemented, examining failures related to product liability that could evoke unwanted crises, such as casualties or recalls.

Let's consider how to proceed once reliability targets are determined. If the lifetime target is set at Y years and the cumulative failure rate to lifetime is X percent for the assembled product, the target for the assembled product is BX life Y years. The target average annual failure rate can be calculated by dividing the cumulative lifetime failure by the lifetime, or X/Y.

Let's discuss how to make a parametric ALT plan for an electric home appliance that has no initial failures and has passed its reliability marginal tests. Assume that the B10 life is 10 years, and therefore the average annual failure rate is below 1%.

---

[*] It is dangerous to guarantee 1 year. Since accidents that occur several years after product release may still lead to bankruptcy, failure rates should be reviewed as well as failure modes that could be expected to occur during up to at least 70% of the product's lifetime.

The final product is disassembled into subassemblies, since breaking it down into individual components is much too complicated. Connecting parts are included with related units. If it is assumed that the number of units is 10, the B1 life of each unit will be 10 years, since the unit lifetime exceeds 10 years.* The average annual failure rate of the units will be 0.1%, which is the core datum for the target setting.

Target setting is performed according to the part-count method, although it is better to set targets with actual field data. Since there are no field data for a new design, the data for similar units are investigated and adapted. If major design changes have been made in structure or material, the failure modes will change and the possibility of increased failure rates in the newly designed units will be high. Since the failure rates of completely new units can sometimes be 10 times higher than those of old ones, rates must be estimated based on the details of the design changes in the component units. In other words, the predicted rates will be X times higher than field records, based on such factors as whether

- The design of the reviewed units remains the same
- The units are only being manufactured by a new supplier
- The load applied to the units is greater than that affecting existing units
- The units are newly designed to produce more capability with the same structure
- The redesign includes proven (or unproven) new technology
- The units are totally redesigned with brand new concepts

The targets for each unit are set using predicted rates. Finally, we sum up all the failure rates of the constituent units, then adjust and fit the sum to meet the target for the finished product, which becomes the completed plan.

Table 5.1 shows the overall plan of parametric ALT for assembled product R-Set, with a targeted B13.2 life 12 years, or an average annual failure rate of 1.1% and a lifetime of 12 years. The estimated failure rates of each unit are established by considering the range of design changes, and the targets for the units are set based on adjusted field data and component size. The newly formulated results tested with parametric ALT must be reviewed to confirm that they meet the targets in the overall parametric ALT plan for the final product R-Set. Of course, if some units have been tested in other models and turned out to meet the target, or if the field data reported by users meet the target, such units need not be tested and their data can be put directly into the plan.

Here the word *unit* means a subassembly or module that has input/output terminals that can be controlled by some outside apparatus. As

---

* See Chapter 3, Section 3.4.

Table 5.1 Overall Plan of Parametric ALT of R-Set (BX Life)

No.	Unit	Main function	Field data		Design change	Transfer	Estimated[a]	Target		
			Annual average failure rate	BX life[b]			Annual average failure rate	Annual average failure rate	BX life	Ref. BX
1	M.P.	Performance control	0.34	5.3	New	×5	1.70	0.15	12	B1.8
2	R.C.	Refrigerant compression	0.35	5.1	—	×1	0.35	0.15	12	B1.8
3	I.M.	Ice making	0.25	4.8	Motor change	×2	0.50	0.10	12	B1.2
4	U.A	For A	0.20	6.0	Partial change	×2	0.40	0.10	12	B1.2
5	U.B	For B	0.15	8.0	—	×1	0.15	0.10	12	B1.2
6	Others	—	0.50	12.0	—	×1	0.50	0.50	12	B6.0
Sum	R-Set	—	1.79	7.4	—	—	3.60	1.10	12	B13.2

[a] Setting up test methods to estimated data would save the expense of initial testing while final test specifications are being established to meet the targets.

[b] Lifetime can be calculated if X of the target lifetime of the unit is divided by the average annual failure rate (in the first column of Table 5.1, e.g., 1.8/0.34 = 5.3 years, 1.8/0.15 = 12 years).

diverse test conditions could easily be considered because of the controllability of input/output powers, testing individual units forces us to select appropriate test methods adequate to the usage conditions. If we test only the final product, the test time to identify failure becomes longer and test expenses are much higher because the severity of the test conditions for the assembled product would be less than that for units. If we use test methods executed with individual components, the sample size would be greater than for unit tests, since the target for the failure rate is far below those for units. The test expenses will be higher than for unit tests, because the working components of the apparatus of input/output terminals require complicated testing equipment. Thus, unit testing is recommended in terms of both the time and expenses required.

Complex finished products, like cars and aircraft, are also aggregates of units—a few hundred in cars and several thousand in aircraft. Therefore, an overall plan for them will include dozens of sheets of parametric ALT for cars and hundreds of sheets for aircraft, the processing for which is basically the same. Units tested in the plan are best arranged according to the order of magnitude of their presumed failure rates—without omissions or a dismissive notation as "chronic problems," because such units can be managed enough to address flaws once they are approached through an adequate procedure.

As noted earlier, two times the annual after-sales service rate becomes the annual failure rate. The rate divided by the annual hours of operation is the hourly failure rate, which becomes the basic input for parametric ALT. Note that sometimes operation time includes nonoperation hours, since chemical changes can still proceed just during those periods.[*]

Although the overall plan of parametric ALT is the test plan for BX life, we can also identify some of the causes of random failure when the tests are performed, and this can reduce failure rates within lifetime. But in order to sufficiently reduce random failure or to reduce it far below X percent of the BX life, other efforts are needed.[†] Generally, to lower random failure, the two elements of stress and materials must be scrutinized; therefore, test conditions responding to peculiar stresses should be used to reveal minor defects and degradation in materials. For this, it is good to make an additional parametric ALT plan for the annual failure rate of the unit. Since it is difficult to approve a product with a failure rate target assumed to be zero when there are no reference data, it is better to set the target as half of the average annual failure rate (X/2), which is easy to apply.

Companies should be careful when setting targets concerning failures related to safety; these should be completely eliminated, regardless of the

---

[*] See Chapter 4, Section 4.4.
[†] See Section 5.7.

effort involved, both because manufacturers generally must reimburse consumers for any casualties once the cause of accidents is identified as lying in the product and because the reputation of the manufacturer suffers, regardless of whether the product exceeds its expected use period. Company M in Japan experienced difficulty after several deaths in 2005 due to forced-draft flue stoves over 12 years old—a case that shows clearly the unlimited liability of a manufacturer.[*]

Some random failures within lifetime, as well as wear-out failures, could become product liability issues. There is a limit on the cumulative failure rate of the BX life in the case of wear-out failure and a limit on the annual failure rate in the case of random failure. How should the targets for such rates be set? The target of the cumulative failure rate (x) and of the annual failure rate (λ) of a certain failure related to PL is calculated by transforming Equation (5.2) as follows:

$$x = \frac{P}{E} \cdot a, \ \lambda = \frac{P}{E} \cdot \frac{a}{L} \tag{5.4}$$

where P is unit price, E is average expenses per case to handle accidents related to the failure, a is the ratio of handling expenses to sales revenue, and L is the product lifetime in years. We can see that failure rates decrease in inverse proportion to the amount of handling expenses/reimbursements. Note that all expenditures related to accidents caused by failures—regardless of whether they occur over several years, even up to product lifetime—should be deducted from the profit of the year when the products were sold.

5.3   Recurrences of the failure mode

For over 20 years the following belief has been treated as a rule in Japan: "If engineers cannot reproduce in the lab the failures that occur in the field, the failure issues cannot be solved." However, often corrective actions failed because they proceeded from rational but not scientific presumptions about the recurrence of failure—in other words, they could not be duplicated away from field use. Recreating failure modes at the factory is sometimes very difficult. The frequent appearance of the terms *no defect found (NDF)* or *retest OK (RTOK)* in reports testifies to this difficulty. In order to produce a recurrence of the fault, engineers must completely understand the situation in which the failure occurs; corrective actions can be verified only by

[*] T. Nishi, "Safety Analysis of Forced-Draft Balanced Flue Stove," *Proceedings of the 38th Reliability and Maintainability Symposium,* Union of Japanese Scientists and Engineers, 2008, pp. 179–184.

testing under the same conditions.* Once a failure has been reproduced, corrective actions can easily be established and confirmed. By the same token, in the development of new products, if dormant failure modes are identified in advance of product release, fixes can easily be implemented. Thus, test methods for discovering potential failures before they occur are very important and must be among the test specifications.

The procedure for improving reliability through actual tests includes the following steps:

1. Search for potential failure modes by testing under various combinations of real environmental and operational conditions. This differs from inspections or assessing whether a measured value conforms to predetermined specifications in the design drawings, because these tests are designed to identify changes in or degradations of the structures of the product. Failed samples can confirm the conditions under which materials will be destroyed or degraded, as well as what kinds of stresses the product experiences and how they operate, combine, and accumulate.
2. Determine the failure mechanisms by analyzing failed samples and clarifying the root causes. For failed samples, confirm that the results are the same as those that would occur in the field.
3. Create alternatives to fix the failures; confirm via tests under the above conditions whether they are appropriate before choosing them as solutions.

These three tasks are not easy. Particularly, identifying the failure mode is very difficult. Irreversible breakage due to large stresses is comparatively easy to find. But the accumulation of stresses on materials, eventually leading to breakage, is hard to identify because it requires a detailed understanding of how manufacturing processes affect materials, of the inherent defects or weaknesses in materials, and again, of the environmental and operational conditions the materials and structures will encounter.

Let's consider step by step why it is virtually impossible to re-create failures in the company lab. If the test conditions differ from the actual field environments where the failures occur, the failure mode cannot be identified. And if the duration of the test does not exactly duplicate the time required for the material to weaken under accumulated stresses, the failure mode will not surface. Finally, the failure mode cannot be replicated if the sample size is relatively small while the failure rate targeted for confirmation is low. Most NDF results do not indicate that the product is faultless; rather, they only show that reliability guarantees are not based

---

* In this sense, establishing test methods based on the no-failure acceptance rule should be approached cautiously, as it is difficult to verify them through actual test results (see Section 5.6).

on an understanding of material behaviors under the relevant conditions and the appropriate statistics. To accurately identify failures, only a thorough grasp of the physical and chemical conditions, as well as the mathematical conditions, will be successful.

Using common sense, let's consider the recurrence test methods used for a failure that occurs at a rate of about 1% in 2,000 h in subtropical conditions. In the subtropics, the main stresses identified are climatic temperature and electric voltage. So, to begin with, if higher temperature and voltage are not imposed on the test items, the failure cannot be replicated. If testing is completed 1,000 h, again the failure cannot be identified, no matter how many samples are tested, because only imposing comparable stresses for over 2,000 h would weaken the materials. And if 10 samples are tested under comparable stresses for 2,000 h, no failures will occur, because only 0.1 of one item among 10 items fail—1 in 100—and thus no failure is likely to occur in a sample size only one-tenth the amount needed to find a failure. It is simple common sense that at least 100 items should be tested under higher temperatures and voltages for 2,000 h in order to produce one failed sample. Higher temperature and voltage are the physical-chemical conditions incurring material degradation and 100 pieces and 2,000 h are the mathematical conditions. In order to confirm a 1% failure rate with a commonsense level of confidence, or about 60%, the test requires 100 samples being tested for 2,000 h.

As this case illustrates, identifying a failure mode under near-normal conditions would take many samples and a lot of time, with consequent great expense. Cutting this requirement down is the task of reliability guarantees. We will explain in turn the outline of parametric ALT (quantitative accelerated life testing), how to reduce test time, and finally, how to cut sample size. In effect, testing 100 pieces for 2,000 h to simulate a subtropical environment can be converted to testing 100 for 150 h, and finally can be reduced to testing 25 pieces for 300 h, as shown in Table 5.2. The

Table 5.2 Outline of Parametric ALT

	Real environment	P-ALT	Reduction of test period	Cutting sample size
Abstract	1% failure in 2,000 h in subtropics	Outline explanation	Calculation of minimum test period with severe conditions	Calculation of sample size reduced in intensive region of occurring failure
Replication condition	100 pieces for 2,000 h		100 pieces for 150 h	25 pieces for 300 h
Explanation	Section 5.3	Section 5.4	Section 5.5	Section 5.6

bases of this method will be explained in terms of simple mental arithmetic in subsequent sections.

5.4   Definition and procedure of parametric ALT

Parametric ALT is an accelerated life test with parameters for quantification. Parametric ALT is defined as a quantitative accelerated life test method that can predict item lifetime, and failure rate within lifetime, using the results of planned testing with an acceleration factor (AF).[*] The AF is calculated with parameters taken from the life-stress model of the dominant failure mechanism occurring in the item (under the given physical and chemical conditions). The minimum test period is determined using the targeted item lifetime and the AF. Sample size is calculated from a Weibull distribution shape parameter, the scheduled test period extended from the minimum period, and the targeted item failure rate (mathematical conditions). To establish a life-stress model that fits the failure mechanism, a knowledge of physics of failure (PoF) is needed; after all, parametric ALT is a scientific approach.

As the AF is the most important factor in minimizing test expenses, it must be carefully determined, using the anticipated physical and chemical conditions the items will experience. The following procedures are recommended for obtaining these with some certainty. For each item tested using the overall plan of parametric ALT for product R-Set, as in Table 5.1, the physical-chemical conditions can be inferred according to a four-step extraction program (4-SEP). This is accomplished by applying the method of two-stage quality function deployment (2-Stage QFD) to the 4-SEP as follows:[†]

1. Survey the structures and materials of the item. Determine how the environmental and operational stresses on the item disperse through its various structures and how they affect the materials of the structures.
2. List all the potential failure mechanisms, based on the relationships between the anticipated stresses and the structural materials.
3. Determine which potential failure mechanisms could become dominant failure mechanisms under the projected environmental and operational conditions (first-stage QFD).
4. Construct a test method, with appropriate stressing conditions, to identify the dominant failure mechanisms (second-stage QFD).

At this point, the AF could be determined from both the actual stresses and the stresses of the test conditions identified with the dominant failure

[*] See Chapter 3, Section 3.5.
[†] J. Evans, *Product Integrity and Reliability in Design*, London: Springer-Verlag, 2001, p. 15.

mechanisms, as explained in the next section. The method of reliability marginal testing discussed in Chapter 4, Section 4.4, can be considered similarly.

Now let's broaden our understanding of the failure modes based on the two elements of failure mechanics.* First, the materials at the failure sites affected by the relevant stresses must be considered. Some common materials, from best to worst reliability, are bronze, steel, aluminum, plastics, rubber, and adhesives. Since the material is a critical factor in reliability, these materials should be thoroughly understood beforehand. This includes assessing the kinds of heat treatments and surface finishes that can cause different behaviors in them. Potential defects due to manufacturing processes and material changes caused by moisture penetration must also be considered.

Second, the stresses at failure sites are concentrated or distributed according to the relevant structures. These should be examined carefully. Reliability issues may exist in awkward structures that are not familiar to our eyes. Natural, balanced structures would be good for long, failue-free performances. Attention should also be paid to structural elements, such as the radii of the curvature of corners and the joints between heterogeneous materials. Of course, the stresses applied to the structure may be thermal and chemical, as well as electrical and mechanical. Structural paths where hot air flows out, or where moisture penetrates, should be examined carefully.

A reliability engineer must study the various failure mechanisms that would be expected to occur at presumed failure sites. Failure mechanisms are divided into two categories: overstress failure mechanisms and wear-out failure mechanisms.† Because there are diverse materials in electronic devices, there are many different failure mechanisms. They can be understood in commonsense terms. Take the failure mechanism called *electromigration*, for instance. This failure phenomenon results when a lot of current flows in the minute patterns of microdevices. Electrons push the metal atoms in the material's pattern, resulting in holes inside and in hillocks near the site, which eventually create open circuits as the holes enlarge, just as vehicle tires push the soil on dirt roads, resulting in ruts and bumps. Another failure mechanism, popcorning, is caused by the alteration in synthetic resins surrounding the microchips as moisture in the resins is vaporized by elevated temperatures, as happens when popping corn. Over 50 such failure mechanisms have been catalogued, so

---

* See also Chapter 2, Section 2.5.
† See also Chapter 3, Section 3.2, and Agency of Technology and Standards. *Briefs of Reliability Terminology*. Gwacheon, Korea: Ministry of Commerce, Industry, and Energy, 2006, p. 247. On the details of failure mechanisms, refer to footnotes * and † on p. 129 in Chapter 6, Section 6.2.

studying them in advance will help in identifying the relevant physical-chemical conditions before test planning.

Implementing parametric ALT involves multiple steps:

1. Carefully establish the targets for lifetime and cumulative failure rate within lifetime, and determine the failure mode for the item.
2. Determine the physical-chemical conditions according to 4-SEP.
3. Create a life-stress model and its equation,[*] using the determined conditions and parameters based on previously acquired data, if available, and estimated conservatively if not. Determine the AF using Equation (5.5) and calculate the minimum test periods required by dividing the target lifetime by the AF.
4. Determine or presume a shape parameter indicating the intensity of wear-out failure, using the Weibull distribution the item failure follows, and then calculate the necessary sample size considering the lifetime target, longer than minimum actual test periods, and the shape parameter, using Equation (5.9).
5. Decide on the appropriate inspection interval to confirm failure during testing. Configure the test equipment and test samples. Preferably, only samples should be inside a severe environmental chamber, with their controllers outside, to prevent damage to the controllers.
6. After finishing the test, calculate the lifetime and confirm that it meets the targets. Document the test bases and results.

Now let's look at two key procedures: constructing a life-stress model for reducing the test period (step 3) and cutting down the sample size (step 4).

5.5   Life-stress model and minimum test period

Multiplying the test hours by the sample number gives the total component hours, as mentioned in explaining the revised definition of failure rate and characteristic life.[†] It might seem to be sufficient to fulfill the total component hours in testing, since it seems that sample size decreases as test hours increase and vice versa. This is true in the case of overstress or random failures, though with some cautions. But the concept of total component hours is completely inapplicable in the case of wear-out failure over lifetime, because test periods to identify wear-out failure are decided using the target lifetime and the magnitude of the stresses under test conditions.

Again, let's consider this by applying common sense. In order to confirm 10 years of lifetime, the test period must be over 10 years; if failures

---

[*] This will be explained in the next section as Equation (5.6).
[†] See Chapter 3, Section 3.5.

are expected to occur epidemically in 3 years, samples must be tested for over 3 years in order to identify the failures. However, because wear-out failures occur as materials are degraded in response to outside stresses, such failures never occur within a certain time period, but only when the materials degrade and weaken to the point that they fail. The reliability engineer should review whether the rates of material degradation are small enough for the material to endure the relevant stresses.

It is meaningless to test for wear-out within minimum test periods because the materials degrade too little in a short test period and would, of course, endure the given stresses. However, as we have noted, conducting tests for longer periods than those of competitors could damage a new product's ability to gain and keep market share. Therefore, the main obstacle to recreating wear-out failures during testing is the number of test hours. In order to overcome this issue, the reliability engineer who wants to predict future failures should test the items under heightened, severe conditions. The more severe the conditions, the shorter the test periods can be; this is the point of an accelerated life test. Thus, minimum test periods can only be decided based on the relationship of the target lifetime and the severity of stresses. In confirming failure rates due to material degradation, it is best to test as close as possible to the lifetime. If the degraded materials receive random stresses, the item should be tested until the materials sufficiently weaken, revealing the failure mechanisms.

Severe conditions speed up degradation and wear-out of materials, so testing under higher stresses than normal is necessary. Higher stresses damage materials faster and accumulate to produce failure faster. However, if testing stresses are increased too much, they can produce different failures than those occurring in actual use. In other words, the real-world failure mechanisms can be overridden by others. For example, the severe stress could cause rupture before the natural cause of degradation was replicated. The failure mechanism should appear to occur in the same way as it does in the field under normal conditions.[*]

In that case, the shape parameter of the failure distribution under severe conditions will be the same as that under normal conditions. When the probability is plotted in a Weibull distribution, the slopes of the two lines under severe conditions and normal conditions are the same, which is called *uniform acceleration*. Because smaller stresses delay the accumulation of material degradation and therefore test periods must be lengthened, the severity of the test conditions should be maximized, both to replicate the wear-out failure as closely as possible and to save expense. One remarkable methodology tests the fly-back transformer, a key unit of television sets with an expected lifetime of over 10 years. The transformer

---

[*] D. Park, J. Park, et al., *Reliability Engineering*, Seoul, Korea: Korea National Open University Press, 2005, p. 283.

is separated from the television set with an extension cord, put into a test chamber, and tested under high temperature. The TV, without the transformer, is located outside the chamber at room temperature and powered on. Failures can be confirmed by looking at the TV screen. This test can identify failure sites within 2 h, resulting in an acceleration factor (AF) of 10,000 (a lifetime of 10 years equals 20,000 h).

Now let's consider the acceleration factor. Typically, acceleration consists of temperature acceleration and stress acceleration, and the value of the acceleration is the product of the two. Temperature rises energize materials and can quickly transform materials into something else. High temperature, even when applied mainly from the outside, can elevate the internal temperature of structural materials and makes them susceptible to molecular alteration. Outside stresses expedite a one-way reaction that overwhelms the normal two-way, balanced reactions.

It is generally accepted that a 10° rise in temperature can double reaction rate, cutting lifetime in half. The quantified value of the change in the reaction rate due to temperature is called the *activation energy*—around 0.6 electron volts (eV).[*] If the activation energy is greater than this, the reaction rate will be faster when there is a 10° rise in temperature. If the stress increases, the lifetime generally would decrease in proportion to the square of the stress ratio. Note that it becomes a proportion of the square, not a simple proportion. We might presume that most stresses in electrical items result from fluctuations in current or voltage and in mechanical items from pressure or physical stress, and thus that the square of their value would be in proportion to energy consumption and the reaction rate would be in proportion to this energy consumption rate. This proportion of the square, or the second power, is generally applicable, but there are many cases where the proportion is higher. For example, the lifetime of a bearing decreases in proportion to the third power of stress[†] and the lifetime related to fatigue in steel decreases in proportion to over the tenth power of stress.[‡]

---

[*] An energy of 5eV is necessary to separate one electron from a natrium atom, and the typical energy interval between the ground state and the excited state is 3 eV. The activation energies of chemical reactions vary from 1 eV to 4 eV (P. Atkins, *Physical Chemistry*, 6th ed., Paju, Korea: Chungmoongak, 2002, pp. 5, 9, 1039). The activation energies of degradation models in semiconductor microdevices are different from failure mechanisms and vary from 0.3 eV to 1.6 eV (D. Park, J. Park, et al., *Reliability Engineering*, Seoul, Korea: Korea National Open University Press, 2005, p. 271). Note that the energy of visible light is 1.8 eV to 3.0 eV and the energy of ultraviolet at 230 nm is 5.4 eV. For reference, the energy necessary for nuclear fission goes up to mega-eV (A. Beiser, *Concepts of Modern Physics*, Seoul, Korea: Haksul Intelligence, 2003, p. 576).

[†] W. Nelson, *Accelerated Testing*, Hoboken, NJ: John Wiley & Sons, 2004, p. 86.

[‡] J. Bannantine, *Fundamentals of Metal Fatigue Analysis*, Englewood Cliffs, NJ: Prentice-Hall, 1990, p. 5.

This concept of the acceleration factor is expressed as follows:[*]

$$AF = \left(\frac{S}{S_0}\right)^n \cdot \exp\left[\frac{E_a}{k}\left(\frac{1}{T_0} - \frac{1}{T}\right)\right] \tag{5.5}$$

where S_0, T_0 are the stress and temperature under normal use conditions, S, T are the stress and temperature under severe test conditions, n is the exponent of stress, E_a is the activation energy, and k is the Boltzmann constant. If the exponent (n) and activation energy (E_a) are found, the AF can be calculated. The lifetime, or anticipated time to failure, divided by the AF becomes the requisite minimum test period to identify failures. The actual test hours can obviously not be less than this minimum test period.

Equation (5.5) can be derived from the equation for the life-stress model (Time-to-Failure Equation).[†] In the life-stress model, the principal stresses can be researched using the 4-SEP method; its parameters— exponent (n) and activation energy (E_a)—should be confirmed. This equation should be easily understood by CEOs, though the reliability personnel would handle it in practice:

$$TF = A(S)^{-n} \cdot \exp\left(\frac{E_a}{kT}\right) \tag{5.6}$$

where TF is time to failure, S, T are the stress and temperature under use conditions, and A is a constant.

Equation (5.6) was derived by separating the stress term related to the lowering of the energy barrier due to outside stress from the temperature term, or the Boltzmann term—the result of a brilliant insight by Dr. J. W. McPherson, of Texas Instruments, in the United States.[‡] Although there

---

[*] J. McPherson, *Reliability Physics and Engineering: Time-To-Failure Modeling*, New York: Springer, 2010, p. 110..

[†] This equation is derived from the following equation (J. McPherson, "Accelerated Testing." *Electronic Materials Handbook. Vol. 1. Packaging*, Materials Park, OH: ASM International, p. 888):

$$TF = A\left[\sinh(aS)\right]^{-1} \cdot \exp\left(\frac{E_a}{kT}\right)$$

The first term [sinh(aS)]$^{-1}$ of the above equation can be substituted with the power term $(S)^{-n}$ or exponential term exp($-aS$) and Equation (5.6) is substituted with the former term. If time to failure changes exponentially, the higher the stress, the more rapidly the degradation of such mechanisms proceeds. As this type of mechanism is scarce (for example, time-dependent dielectric breakdown (TDDB)), most failure mechanisms can be arranged as power terms.

[‡] J.W. McPherson, "Stress Dependent Activation Energy," *Proceedings of the 24th International Reliability Physics Symposium*, IEEE, 1986, p. 12.

are exceptions, the AF can be divided into factors related to stress and temperature. The parameters (exponent (n) and activation energy (E_a)) can also be estimated intuitively from related experience, which means that the life-stress model can be understood straightforwardly. Because this equation evolved from the model of microdepletion or microaccumulation of materials occurring in the failure region, it is applicable to almost the whole range of failure phenomena in mechanical and electronic items.[*] Therefore, this equation, with its two terms, becomes the generalized life-stress equation.[†] The first of the two terms is the outside stress term and the latter is the internal energy term. Using physical-chemical terminology again to explain this, the outside stress lowers the energy barrier and the high temperature activates the elements in the material; the highly energized elements easily cross over the barrier, which means that materials gradually degrade and finally fail.[‡] Here we might add that the thermal shock (ΔT) frequently applied in testing is not the internal energy term, but an outside stress term, because the thermal shock induces strain change, which affects the ability of materials to accommodate stresses.

Equation (5.6) has three unknown values—constant A and two parameters. Useful data have been reported in the proceedings of many symposiums that would be applicable to item testing. In some cases the exponent increases[§] and the activation energy decreases[¶] according to the rise in stress; this can be determined by surveying failure mechanisms and test conditions.

In cases when there are no data or when the data need to be confirmed, these parameters can be found using a three-level test. Again, note that the failure mechanisms identified by the test results should be identical. The critical failure mechanisms and their life-stress models are clearly described in the materials by McPherson.[**]

If the one constant and two parameters of Equation (5.6) are found, the lifetime for that failure site can be calculated approximately. But the BX life cannot be calculated, because no distributions related to structure and material characteristics are included. If constant A is determined with the variables of structure and material and the distributions are substituted

[*] J. McPherson, "Accelerated Testing," *Electronic Materials Handbook*, Vol. 1, *Packaging*, Materials Park, OH: ASM International, p. 888.

[†] D. Ryu, "Novel Concepts for Reliability Technology," *Proceedings of the 36th Reliability and Maintainability Symposium*, Union of Japanese Scientists and Engineers, 2006, p. 34.

[‡] A. Grove, *Physics and Technology of Semiconductor Devices*, Hoboken, NJ: John Wiley & Sons, 1967, p. 36.

[§] J.W. McPherson, "Stress Dependent Activation Energy," *Proceedings of the 24th International Reliability Physics Symposium*, IEEE, 1986, p. 12.

[¶] J. McPherson and D.H. Baglee, "Acceleration Factors for Thin Oxide Breakdown," *Journal of the Electrochemical Society*, 132(8), 1985, 1903.

[**] J. McPherson, "Accelerated Testing," *Electronic Materials Handbook*, Vol. 1, *Packaging*, Materials Park, OH: ASM International, pp. 889–893 and *Reliability Physics and Engineering: Time-To-Failure Modeling*, New York: Springer, 2010, pp. 137–267.

for their variables, the BX lives responding to specific stresses can be calculated.[*]

Let's calculate the AF in the case given in Section 5.3. Here, the main stresses are temperature and voltage. If the sets operating at 30°C (on average) were tested at 60°, failures would increase eight times, or triple two times. The temperature rise is 30°, and the acceleration rate of two times per every 10° repeats three times ($2 \times 2 \times 2 = 8$). If the sets operated mainly at 220 V were tested at 286 V, which is an increase of 30%, failures occur 1.7 times faster, which is the square of 1.3. The acceleration factor for these conditions (temperature at 60° and high voltage at 286 V) can be calculated by multiplying the two—8 × 1.7—which gives an AF of 13.6 times. Two thousand hours of testing under normal conditions can thus be reduced to 2,000/13.6, or a 150 h, under severe conditions. This speeds up the test of 100 pieces for 2,000 h in a subtropical environment to an equivalent test of 100 pieces for 150 h under conditions of increased severity.

As explained in Chapter 3, we cannot anticipate the behavior of an item after the period for which it has been tested. In this example, the results within 2,000 h of operation can be predicted, since the test of 150 h under severe conditions produces an equivalent duration. Therefore, test hours cannot be reduced arbitrarily. If there were no way to induce acceleration factors, items would have to be tested until the target lifetime under normal conditions, which would seldom be feasible. Fortunately, there are always severe conditions and other methods that can be used to reduce test periods. For example, after the number of stress incidents that would occur up to the target lifetime is determined, the test period for an item could be lessened by decreasing the idle time between stresses.

In actual environments, materials are more likely to encounter two or more stresses than just one, and these materials obviously weaken more easily than those experiencing only one stress. The multiple-stress environment can easily be adapted using the AF equation (Equation (5.5)). For example, using this equation the period during which two levels of stress are imposed can be transformed into the equivalent period simulating only one level of stress. In the case of an automobile, the comparison between use in tropical areas with bumpy roads and temperate regions with well-built expressways can be expressed quantitatively with this equation. Consequently, the reduction of vehicle lifetime and the increase in annual failure rate in the tropics vs. in temperate regions can be predicted; the warranty cost can also be estimated.

The synthesis of stresses known as Miner's rule pertains when there is only one kind of stress, but at different levels. Therefore, this rule cannot

[*] J. Evans and C. Shin, "Monte Carlo Simulation of BGA Failure Distributions for Virtual Qualification," *Advances in Electronic Packaging*, 2, 1999, 1191–1201.

be applied to units used in environments with two or more stresses, such as steel components with bearings that have lubricants or greased rubber seals. The AF equation can be applied to this kind of situation.[*] For example, if the number of cycles of mixed stresses is confirmed via field data, the following equation can calculate the number of cycles of the specific stress (n_0)—that is, the greatest stress in the field—as follows:[†]

$$n_0 = \frac{n_1}{AF_1} + \frac{n_2}{AF_2} + \cdots \qquad (5.7)$$

where AF_1, AF_2 are acceleration factors of imposed stresses to the specific stress, and n_1, n_2 are the numbers of cycles for the imposed stresses.

The number of imposed cycles for the specific stress becomes the target cycle number of the test plan. The test planner must calculate the acceleration factor to create severe conditions, which will of course include higher stresses than the equivalent specific stress the item receives in normal use.

The shortened test period could have been determined with the AF. The sample size—the final stage of test specification design—will be discussed in the next section.

5.6   Shape parameter and sample size reduction

In reliability test planning, the determination of sample size is the last big hurdle. It is relatively easy to calculate the sample size when testing for the failure rate, but more difficult to work out the sample size when establishing the item's lifetime target. Some reliability engineers in advanced corporations can reference a table to do this or use a particular software package that their specialists have worked out in advance. But this process has its drawbacks; if the underlying bases for the figures are not understood, the test will not be carefully calibrated to actual conditions and cannot be modified to accommodate them.

Let's consider the equation for sample size with a commonsense level of confidence. It is sufficient if the estimated lifetime (L_B) drawn from the test is longer than the target lifetime (L_B^*). The estimated lifetime can be derived if the revised definitions of lifetime (Equation (3.2)) is substituted for the characteristic life (η^β) in the BX life equation (Equation (3.3)), and if acceleration factor are put into the test period. As mentioned earlier,[‡]

---

[*] The acceleration factor can be calculated by combining the stress acceleration by the exponent of stress obtained from the S-N curve and temperature acceleration.

[†] D. Ryu, "Novel Concepts for Reliability Technology," *Proceedings of the 36th Reliability and Maintainability Symposium*, Union of Japanese Scientists and Engineers, 2006, p. 37.

[‡] See Chapter 3, Section 3.5.

the variables related to time are modified with the exponent of the shape parameter (β):

$$L_B^{*\beta} \le L_B^{\beta} \cong x \cdot \eta^{\beta} \cong x \cdot \frac{n \cdot h^{\beta}}{r+1} = x \cdot \frac{1}{r+1} \cdot n \cdot (AF \cdot h_a)^{\beta} \tag{5.8}$$

where x is the cumulative failure rate, r is the failed number, AF is the acceleration factor, and h_a is the actual test period, reduced by imposing severe conditions. Solving Equation (5.8) for sample size yields[*]

$$n \ge (r+1) \cdot \frac{1}{x} \cdot \left(\frac{L_B^*}{AF \cdot h_a} \right)^{\beta} \tag{5.9}$$

This equation seems complex, but its concept is simple and easy to memorize if separated term by term, as shown in Table 5.3.

Next, let's calculate the sample size (n) sufficient to establish a failure rate target estimation with a commonsense level of confidence. The test is adequate if the estimated failure rate (λ) drawn from the test is smaller than the target failure rate (λ^*). The estimated failure rate can be derived if the acceleration factor is put into the term for the test period in the revised definition of failure rate (Equation (3.1)):

$$\lambda^* \ge \lambda \cong \frac{r+1}{n \cdot h} = \frac{r+1}{n \cdot (AF \cdot h_a)} \tag{5.10}$$

Solving Equation (5.10) for sample size yields

$$n \ge (r+1) \cdot \frac{1}{\lambda^*} \cdot \frac{1}{AF \cdot h_a} \tag{5.11}$$

---

[*] The approximate equation (Equation (5.9)) can be derived from the following equation, on condition that the cumulative failure rate is sufficiently small and the confidence level is about 60% (commonsense level):

$$n \ge \frac{\chi_\alpha^2 (2r+2)}{2} \cdot \frac{1}{\ln(1-x)^{-1}} \cdot \left(\frac{L_B^*}{AF \cdot h_a} \right)^{\beta} + r$$

$$\cong (r+1) \cdot \frac{1}{\left(x + \frac{x^2}{2} + \frac{x^3}{3} + \cdots \right)} \cdot \left(\frac{L_B^*}{AF \cdot h_a} \right)^{\beta} + r \cong (r+1) \cdot \frac{1}{x} \cdot \left(\frac{L_B^*}{AF \cdot h_a} \right)^{\beta} + r$$

The derivation of this equation can be found in D. Kim, "The Study of Reliability Assurance System," doctoral dissertation, Sungkyunkwan University, 2006, p. 47, and D. Ryu, "Novel Concepts for Reliability Technology," *Proceedings of the 36th Reliability and Maintainability Symposium*, Union of Japanese Scientists and Engineers, 2006, pp. 28, 31.

Table 5.3 Summary of Parametric ALT (Commonsense Level of Confidence)

	Real environment	Reduction of test period	Cutting sample size
Target/basis	1% failure in 2,000 h in subtropics	AF = AF(voltage)*AF (temp) = 1.7*8 = 13.6	Failure intensity (β) = 2 Test period increase = 2 × (150 h) = (300 h)
Interpretation/ application	$n \geq (r+1) \cdot \dfrac{1}{x}$	$h_a = h/AF$ $= 2000/13.6$ $= 150$ h (minimum test period)	$R = \left(L^*/AF \cdot h_a\right)^{\beta}$ $= \left(2000/(13.6 \times 300)\right)^2$ $= \left(1/2\right)^2 = 1/4$
Test specifications I	100 pieces for 2,000 h in subtropics with no failure found	100 pieces for 150 h in severe conditions with no failure found	25 pieces for 300 h in severe conditions with no failure found
Test specifications II	200 pieces for 2,000 h in subtropics with one failure found	200 pieces for 150 h in severe conditions with one failure found	50 pieces for 300 h in severe conditions with one failure found

Let's consider the meaning of the two equations for sample size, Equations 5.9 and 5.11. Multiplying the numerator and denominator of Equation (5.11) by the lifetime or target period (L^*) yields

$$n \geq (r+1) \cdot \frac{1}{\lambda^* \cdot L^*} \cdot \frac{L^*}{AF \cdot h_a} = (r+1) \cdot \frac{1}{x} \cdot \left(\frac{L^*}{AF \cdot h_a}\right)^1 \tag{5.12}$$

Here, $\lambda^* \cdot L^*$ becomes the cumulative failure rate (x), as it is the result of multiplying the failure rate per period by the specified period (for example, lifetime).

Comparing the sample size equations for failure rate (Equation (5.12)) and for lifetime (Equation (5.9)) implies that they have a similar form. The difference between the two equations is whether the exponent of the third term is 1 or β—which is greater than 1 for wear-out failure. Therefore, the sample size equation for the failure rate (Equation (5.12)) can be included in the sample size equation for the lifetime (Equation (5.9)).

Consequently, the sample size equation for the lifetime (Equation (5.9)) becomes a generalized equation for sample size, including failure rate and lifetime, with a commonsense level of confidence:

$$n \geq (r+1) \cdot \frac{1}{x} \cdot \left(\frac{L_B^*}{AF \cdot h_a} \right)^{\beta}$$

This equation constitutes three terms:

(sample size) ≥ (failed number + 1)

    × (1/cumulative failure rate)

    × ((lifetime target)/(AF × actual test period))$^{\beta}$

The first term is related to the number of failures to be found in the test, the second term relates to the failure rate, and the last term is a reducing factor (R) that is the ratio of the lifetime target to the test period. As mentioned earlier,[*] the sample size will be reduced if the denominator is greater than the numerator.

Let's mentally calculate the two principal test conditions, sample size and test period. If we test an item for as long as its targeted lifetime, the reduction factor becomes 1. Since the third term, or reduction factor, equals 1 regardless of the failure intensity or shape parameter, the generalized equation for sample size (Equation (5.9)) will be

$$n \geq (r+1) \cdot \frac{1}{x} \qquad\qquad (5.13)$$

This is just the basic commonsense mental equation for sample size. The number of samples equals the inverse of the cumulative failure rate when no failure is found. For example, assume that the failure rate target is 1% per 2,000 h, or a B1 life of 2,000 h, and that the duration of the test is scheduled to be the same as the target period (2,000 h). If the number of failures identified is zero, the first term of Equation (5.9) becomes 1, with a commonsense level of confidence. If there is no failure in a test of 100 items for 2,000 h, the above target is verified. If the failed number to be precipitated is 1, the first term of Equation (5.9) becomes 2; if the failed number is set at 2, the first term of the equation becomes 3. In these cases, the test specifications that would satisfactorily verify the targets would be different: one or no failures to be found in a test of 200 items for 2,000 h, and two or fewer failures found in a test of 300 items for 2,000 h.

The sample size increases proportionately with the failed number to be found, and increases inversely as the failure rate decreases. In order to raise the confidence level to around 90%, the number of samples must

---

[*] Refer to Chapter 3, Section 3.6.

increase approximately two times over the commonsense level of confidence, or $(2 \times (r + 1))$.

Now let's apply the third term to the sample size calculation. If the test period is longer than the target period, the sample size decreases. For failure rate tests, the number of samples decreases in proportion to the length of the test period; for lifetime tests, the number of samples decreases further, because the third term is the β-powered value of the ratio of the lifetime target over the virtual test period, where the virtual test period is the actual test period multiplied by the AF. For example, when the failure intensity of the wear-out failure, or the shape parameter, is 2 and the virtual test period is scheduled to be two times as long as the lifetime target, the number of samples will decrease not by half, but by one-quarter.

If the wear-out failure concentrates during a certain period, the shape parameter becomes greater than 1; for example, the shape parameters of blades or of disks in an aircraft turbine engine reach around 7 or 8.[*] Many items have a shape parameter over 2; it is not difficult to raise the shape parameter of items from below 2 to over 2 by improving the structure or materials.

In order to identify failure, a test of 100 pieces for 2,000 h is needed in the subtropics,[†] but applying severe conditions changes the specifications for the test to 100 pieces for 150 h (equivalent to about a 2,000 h virtual test period), on condition that no failure is found. If the actual test period were extended to 300 h (about 4,000 h of virtual testing) under the same severe conditions, the number of required samples would decrease to one-quarter of the set of 100—that is, 25 pieces for 300 h. These figures are summarized in Table 5.3. The CEO who understands the logic of these processes has acquired a fundamental knowledge of verification specifications. Actual verification specifications with more complicated arithmetic can be calculated in the same way as the simple mental calculations discussed here.[‡]

So far, we have discussed testing on the condition that no failure is found—that is, the number of failures is zero—for ease of explanation. Now let's consider the concepts of test specifications when the test produces failed items.

If one or two failures are found, as estimated, after accelerated testing, the newly established test specifications have accounted for the physical and chemical conditions well. If there are more than two failures due to the failure mechanisms that were predicted, the result would still be

---

[*] R. Fujimoto, "Lifetime Estimation by Weibull Analysis (Excel)," *Proceedings of the 30th Reliability and Maintainability Symposium,* Union of Japanese Scientists and Engineers, 2000, p. 248.

[†] This is the example in Section 5.3.

[‡] This is an example of using common sense to grasp a complex concept, as mentioned at the beginning of Chapter 3. Difficulty of explaining with complicated equations results from insufficient understanding of the foundational ideas that underlie them.

quite good, indicating that the test specifications were adequate and that the shape parameter can be confirmed. The ability to confirm the failure mechanism means the test is successful. Test specifications should be designed to produce at least one failure. This validates the test and allows approval of the product.

If no failure is found, the product may appear to be reliable, but we have gained no further information about it. The test specifications would have to be considered tentative and followed up with other actions. One solution would be to extend the test period longer than planned. Another would be to test another set of different samples. If, after taking these actions, there is still no failure, the life-stress model should be reviewed again, since the identified stresses are inappropriate. There is always a failure site that limits lifetime, always a weakest spot. Thus, it is dangerous to design test specifications for new items based on confirming nonfailure. When the test uncovers no failures, it is difficult to know whether the item is truly reliable, or whether the test specifications are simply inadequate to confirm reliability.* It is the task of reliability managers to determine whether the lifetime of the weakest site equals the targeted lifetime for the product or not.

At the same time, care has to be taken that the test does not become a self-fulfilling prophecy. When the test period is extended too far, the failure may occur naturally, as intended. But there is also a risk that a failure may surface that is unlikely to occur in the field. If the time to precipitate a failure far exceeds the estimated time to failure, the stresses in the test should be reviewed for conformity to those occurring in use conditions. Again, the reliability engineer should carefully follow the 4-SEP process. That means that the test should be reviewed to determine whether or not the dominant failure mechanisms are omitted, with a careful analysis of the materials in the item and the stresses they experience. Of course, if the structure includes new materials that may produce unanticipated failure mechanisms, it is critical that the test replicate as closely as possible actual field conditions.

5.7   Characteristics of life tests and difficulties with the failure rate test

The sample size equations for lifetime and failure rate follow the same structure as just explained. There is only one difference regarding the shape parameter and the ensuing reduction factor. The test combinations of sample sizes and test periods for different failure rates are presented in Table 5.4. The basic conditions for test specifications are a lifetime target

---

* Recall how difficult it is to identify failure, as discussed in Section 5.3.

Table 5.4 Comparison of Test Conditions According to Approval Target

Approval target (failure rate or BX life)	Failure rate test	Lifetime test		
	Shape parameter = 1	Shape parameter = 2	Shape parameter = 3
1%/2,000 h	100 pcs, 150 h	100 pcs, 150 h	100 pcs, 150 h
B1 life 2,000 h	50 pcs, 300 h	25 pcs, 300 h	13 pcs, 300 h
	30 pcs, 500 h	10 pcs, 500 h	4 pcs, 500 h[a]
0.1%/2,000 h	1,000 pcs, 150 h	1,000 pcs, 150 h	1,000 pcs, 150 h
B0.1 life 2,000 h	500 pcs, 300 h	250 pcs, 300 h	125 pcs, 300 h
	300 pcs, 500 h	92 pcs, 500 h	28 pcs, 500 h
	100 pcs, 1,500 h	—	—
0.2%/2,000 h	500 pcs, 150 h	—	—
	250 pcs, 300 h	—	—

[a] Because, in the process derived from Equation (5.9), the approximation condition from the binomial distribution to the Poisson distribution is included, the error when the calculated sample size is small becomes greater than when the calculated sample size is big. Therefore, in this case, the sample size should be increased by one or two pieces.

of 2,000 h, testing under severe environments with an AF of 13.6, and a goal of zero failures with a commonsense level of confidence. Note that random failure is assumed not to be unusual, but related to the conditions that limit lifetime—for example, defects in materials.

The required condition in Table 5.4 is that the lifetime test period exceeds 2,000 h. This is sufficient because the test periods are over 150 h and the AF is 13.6; thus, the multiplication of the two reaches 2,040 h. Therefore, the test periods of all the combinations are longer than the minimum test period, or 147 h, which is calculated as 2,000 h divided by the AF of 13.6.

Let's take a look at lifetime tests ($\beta = 2$, $\beta = 3$). In order to confirm the cumulative failure rates of 1% and 0.1% in the test for 150 h (no failure found), we can easily calculate that the necessary sample size will be 100 pieces and 1,000 pieces, respectively. Increased test hours decrease the required sample size inversely by the square and the cube of the ratio of test hours, respectively, when the shape parameters are 2 and 3. We can mentally calculate the test combinations of sample size and test period (in the right column of Table 5.4) that meet the targets in the left column.

Now let's consider the characteristics for the lifetime test. Items must be tested for the equivalent of the lifetime. In order to reduce test time, severe conditions are applied. A period of 150 h will exceed the minimum test hours predicted to find failure. As the test period approximates the lifetime, the failure mode should appear once the number of samples is sufficient. If the test period is extended two times, to 300 h, or over three times to 500 h, the materials in the structure will experience the same kind of intensive wear-out and degradation, and thus failure. There would in fact be many failures, assuming that the sample size is the same as required for the minimum test hours. Therefore, the sample size can be reduced drastically because a few failed samples are sufficient to appraise item reliability. This results in economizing test expenses. In addition, the greater the shape parameter, the more drastically the sample size decreases. To save test expenses further, the shape parameter or the severity of test conditions should be increased by reliability engineers.

Parametric ALT of lifetime testing is very helpful for identifying a certain level of random failure. For example, assume that 250 pieces with a shape parameter of two are tested for 300 h with a 13.6 AF in order to verify a B0.1 life 2,000 h, as in line 5 of the third column in Table 5.4. These are the lifetime test specifications, which are the same as for a failure rate test verifying a random failure rate of 0.2% for 2,000 h (line 9 of the second column in Table 5.4). Therefore, the random failure of 0.2% and over will also be identified. This random failure may be found at any time during the testing, not only as the test reaches the vicinity of the lifetime. Thus, the test can also be used to identify the random failure probability of weak materials damaged in manufacturing under similar environmental conditions as the lifetime test.

What are the characteristics of a random failure test? As shown in the second column of Table 5.4, the total component hours, or the multiplication of the sample number and the test period for the same target lifetime, are constant. There is no way of reducing the total component hours without elevating the severity of the test conditions. Only severe conditions can reduce test expenses.

Let's pursue this. In the table, the sample numbers for the same test period—500 h for the two different targets, or 1%/2,000 h and 0.1%/2,000 h—are 30 pieces and 300 pieces, respectively. In order to confirm the latter target, there should be 10 times more effort than for the former target. Conceptually, 10 times the sample number will be necessary for the same test period (line 3 and 6 of the second column in Table 5.4) and 10 times the test length will be necessary for the same number of samples (lines 1 and 7 of the second column). This suggests that action should be directed toward lowering the random failure rate, which originally comes from item characteristics.

Random failure occurs when extraordinary stresses are imposed on materials or when materials damaged by manufacturing processes without screening receive ordinary stresses. The former case indicates special stresses leading to failure—for example, fairly high voltage, or a power surge from the input cable of electrical appliances. For example, engineers trying to confirm these stresses with electrical instruments probably will not find any high-voltage spikes in 1 day, but if the testing period is extended to 10 days, they can perhaps identify at least one instance of high voltage. The latter finding implies the existence of manufacturing flaws that will lead to random failures. These errors do not happen consistently—only accidentally. This means that one mistake occurring in 10,000 operations cannot be checked by observing 1,000 operations.

Let's say more about the reduction of the failure rate, for example, by one-tenth, in terms of two elements of failure mechanics. If the problem is assumed to be with the materials, the manufacturing manager should identify 10 times more critical-to-quality features as quality control items and improve them, since the failure rate target is far less than one-tenth. The whole process should be investigated with the goal of finding problems that can be activated by chance. In this sense, it is impossible to accurately confirm problems with a small sample size of an item made with a large number of processes. Because the failure rate of electronic microdevices is extremely low, the sample size for their approval should increase inversely with the failure rate target. If the problem is assumed to be related to stress, the manager should consider at least 10 times longer periods than those of a previous study. The stress magnitude and its frequency should be characterized in detail. And the possibility of raising the current stress conditions should be considered. This is the same as studying the accumulated stresses on the structure of automobiles delivered by various roads in diverse regions over lengthy periods of operation using sensors attached to particular spots on the cars. The analysis of accumulated stresses may indicate higher stresses than previously assumed, resulting in a decreasing AF and the need to increase actual test hours. If any particularly high stress is identified, the conditions of the reliability marginal test would also change. After the stresses are adjusted, the failure rate test becomes the test to find problems on the material side.

Let's consider the test conditions for a random failure rate of 1% per 2,000 h, as in Table 5.4. Which is better among the three with the same total component hours of 15,000: 100 pieces for 150 h, 50 pieces for 300 h, or 30 pieces for 500 h? Again, if the primary concern is stress, the longer test period would be more appropriate; if the concern is with the materials, it would be better to test larger numbers of samples.

It is not surprising that there are so many steps involved in lowering the random failure rate in advance. This kind of failure cannot be solved casually. Based on the above concepts, reliability engineers can direct their attention to where defects are likely to be located or to how mistakes are overlooked. In order to lower the failure rate from 1% to 0.1%, parametric ALT should be designed in detail, identifying where troubles lie dormant. Once these are located, failure can be addressed with relevant technology and quality control methods. Lowering the random failure rate is difficult to address, but it can be done.

5.8 Redundant systems and preventive maintenance

What should we do when engineers cannot sufficiently lower the failure rate to meet the target? Not lowering the failure rate sufficiently means that it is impossible to perform adequate tests because we cannot precisely identify outside stresses or the state of constituent materials, or because the very low failure rate target needed to provide safety makes performing the appropriate tests too time consuming. At this time additional items should be added to the system or additional systems can be adopted. These are called *redundant systems*. There are, for example, three sets of computers in the Boeing 777 for primary flight control. With the additional two computers, even if the first two malfunction, the plane can still navigate safely.

If we apply redundant systems to a certain item, by how much can we lower the failure rate? If we incorporate a redundant system of two of the same items having an annual failure rate of 1%, the failure rate of the overall system will be decreased by the square of its failure rate, or $(0.01) \times (0.01)$, to an annual failure rate of 0.01%. If the redundant system has three items, its failure rate will be decreased by the cube of the item failure rate, resulting in an annual failure rate of 0.0001%. If an assembled product has around seven thousand of such items, for example, having a hundred of components, each with an annual failure rate of 0.0001%, the failure rate of the product will be 0.7%.[*] The magnitude of this failure rate explains why the Boeing 777 has three computers—triple-triple redundant architecture. Furthermore, each of the computer processors in that system has a different maker.[†] The software has also been provided by different compilers. All these redundancies are planned to avoid common mode failures by using hardware with different structures, because it is so difficult to identify in advance the particular external stresses and microdefects in the materials receiving them.

The functional items influencing safety require redundant systems, though they increase the cost, because all the stresses and the state of the materials cannot be known at a level to match the extremely lowered failure rate target. If all the stresses could be identified, if enough time could be devoted to testing, and if the structures (including materials) could be

---

[*] Assume that other items are not constituted of redundant systems and that their annual failure rates are the failure rate of a triple redundant system, or 0.0001% (see Table 3.3). In many cases, it is hard to attain the target of an annual failure rate of 1% in an item with 1,000 components, though it differs substantially from item to item.

[†] D. Siewiorek and P. Narasimhan, "Fault-Tolerant Architectures for Space and Avionics Applications," Pittsburgh, PA: Carnegie Mellon University, p. 13, available online in PDF format, accessed October 9, 2010.

improved according to the test results, complex products with extremely low acceptable failure rates would not need redundant units, or perhaps would have only one.

What should be done when the item lifetime cannot meet the target, or when the system must be operated long after its predicted lifetime? It is preferable to perform preventive maintenance or to substitute new items for used ones when the BX life confirmed through parametric ALT is reached. Around the vicinity of the BX lifetime items degrade rapidly and are almost certain to fail. Note that if the replacement items are exposed to outside environments, they cannot be used for substitution because they will already have degraded to some extent even if the items are not operated. Thus, replacement items should be preserved in environments with the lowest possible temperatures and humidity. For the same reason, microcircuit devices and remnant devices from the manufacturing process should be stored in tubes preventing moisture from infiltrating.

The exchange interval for units or systems that are standardly used should also be reconsidered. These intervals should be determined in terms of the BX life. Of course, the cumulative failure rate, X, of the BX life will be established uniquely, according to the function and complexity of the item.

In order to extend the system lifetime when the lifetimes of constituent items are comparatively short, it is useless to design redundant systems, because duplicate items incorporated into the system will fail at about the same time. Moreover, if the system is expected to have a comparatively high random failure rate, preventive maintenance cannot reduce the failure rate.* The replacement items would not be improved in relation to particular stresses. These two concepts may seem to run contrary to the previous discussion. As an analogy, consider that in order to reliably preserve a family line, many children should be born when the infant death rate is high, since lowering the death rate may be almost impossible in, say, a less developed country where there are many causes of death. And in order to make an aging man live longer, the replacement of failing organs would be necessary. The parts likely to stop functioning in a man who survived various random accidents could be estimated, and he would live longer if those parts could be replaced.

In brief, when the failure rate within the lifetime is higher than the target, it should be designed with redundancy; if the lifetime is shorter than the target, replacement items should be prepared for preventive maintenance. The potential countermeasures are as disparate as the possible causes of the higher failure rate and shorter lifetime.

* D. Park, J. Baek, et al., *Reliability Engineering*, Seoul, Korea: Korea National Open University Press, 2005, p. 307.

This discussion is appropriate to cases where there is no initial failure and minimum random failure; otherwise, it is meaningless to design redundant systems and stockpile spare parts. Initial failures are easy to recognize and represent major problems.

5.9   Estimating reliability targets from test conditions and predicting expenditures for quality failures

Now let's infer the reliability target of a certain item with a commonsense level of confidence, looking at its test specifications: test period, sample size, and test conditions. Solving Equation (5.9) for the cumulative failure rate (x) yields

$$x \cong (r+1) \cdot \frac{1}{n} \cdot \left(\frac{L_B^*}{AF \cdot h_a} \right)^\beta \qquad (5.14)$$

If the test hours are consistent with the target period, with the condition that no failure is found, the cumulative failure rate target (x) will be the inverse of the sample number and the target period (h) will be the product of the actual test hours (h_a) and the acceleration factor (AF):

$$x \cong \frac{1}{n}, \; h = AF \cdot h_a \qquad (5.15)$$

At this time, the AF would be estimated considering the severity of the test conditions. Therefore, we can recognize that, in the case of the lifetime test, the lifetime target is BX life h hours and, in the case of the failure rate test, the failure rate target is x/h. In other words, the fact that no failure is found in a test of 100 pieces for 2,000 h with an AF of 10 means that the item's reliability is verified with a lifetime target of B1 life 20,000 h, or a failure rate target of 1%/20,000 h. A targeted lifetime that exceeds the test period cannot be accepted.[*] Therefore, the target period will become the minimum assurance period and the failure rate target for no failure found will become the maximum failure rate. Of course, if the assurance period is reduced, the cumulative failure rate target decreases, as expressed in Equations 5.14 and 3.4.

Table 5.5 shows a framework for experiments related to the development of a medicine. Let's infer the quantitative targets and estimate the

---

[*] This is explained in Chapter 3, Section 3.6.

Table 5.5 Clinical Experiments for a New Medicine

	Animal experiments	Clinical trial		
		Phase 1	Phase 2	Phase 3
Experiment goal	Confirming safety and effectiveness	Determining safety and measuring dosage	Confirming medicinal effect and side effects	Reconfirming medicinal effect and long-term safety
Sample size	—	(20)~30 persons	(100)~300 persons	(1,000)~5,000 persons
Time required	3 years	1.5 years	3 years	3 years

Source: Y. Choi, "R&D Strategy for Promoting Bio Industry." Table III-19, Seoul, Korea: Korea Institute for Industrial Economics and Trade, 2006, p. 77.

expense risk due to side effects and so on. Let's also identify issues related to clinical experiments and consider their correctives.

Since the side effects of medicines can be regarded as problems, the side effect rate is equivalent to the concept of the failure rate. If a new medicine evidences no problems in a series of clinical trials, as shown in Table 5.5, its effectiveness can be accepted. However, since phase 3 in the clinical trials has the same test period (3 years) as phase 2, saying that it confirms long-term safety would be incorrect. For a medicine that "passes" in phase 2 or phase 3, usage should be confined to within 3 years under normal conditions, because no severe conditions have been imposed to test for a longer safety period. If there are no side effects in the phase 2 trial of 300 people, the BX life would be B0.33 life 3 years, or a side effect rate of 0.33% for 3 years according to Equation (5.15); the average annual side effect rate would be 0.11% according to Equation (3.1). The conclusion of the phase 3 trial of 5,000 people would be a B0.02 life of 3 years, or a side effect rate of 0.02% for up to 3 years and an average annual side effect rate of 0.0066%. These side effect rates will be the maximum value estimated from the experimental results with a commonsense level of confidence.

Now let's calculate the ratio of the reimbursements resulting from side effects to sales, with and without an accumulation effect. An accumulation effect means that taking medicine or inserting an implant into the human body would slowly aggravate health problems over time by the minute accumulation of damage due to pharmacological action or by a sudden rupture or leakage due to degradation by immune reactions. The latter situation, corresponding to random failure, would be analogous to the concept of failure rate and the former, corresponding to wear-out failure, would involve the concept of BX lifetime.

In the case that there is no accumulation effect, let's calculate the ratio of reimbursements to sales using the results of the phase 3 trial. As mentioned above, the average annual side effect rate (f) is 0.0066%. If the unit price (P) of the new medicine is $0.1 and the average reimbursement (E) is $100 for a person experiencing damaging side effects, the ratio (a) of reimbursements to sales can be estimated according to Equation (5.3) as follows:

$$a = f \cdot \frac{E}{P} = (0.0066\%) \cdot \frac{(100)}{(0.1)} = 6.6\%$$

As the 6.6% of sales revenue would be considered the handling expenses of the side effect, the CEO would recognize the potential for heavy losses and reconsider how many people should be tested and for how long in the clinical trials.

In the former case, in which the damage accumulating due to a medicine or implant (for example, a breast augmentation) is estimated, let's calculate the ratio of reimbursements to sales. Assume that the same experiments are conducted for approving the medicine or device. Although many people are tested in phase 3, the results are insufficient to estimate the ratio because of the comparatively short test period. However, if the 300 people in the phase 2 trial also participated in phase 3, it means, fortunately, that there were no problems for 6 years. However, since the test period is 6 years, the use of the medicine or device should be confined to 6 years. The final results of phases 2 and 3 indicate that the cumulative side effect rate (x) for 6 years is 0.33%, according to Equation (5.15), or B0.33 life 6 years. If the average handling expenses (E) related to the side effects cost $20,000 and the unit price (P) of the medicine or implant is $1,000, the ratio ($a$) of reimbursements to sales can be estimated according to Equation (5.2) as follows:

$$a = x \cdot \frac{E}{P} = (0.33\%) \cdot \frac{(20,000)}{(1,000)} = 6.6\%$$

Because this ratio is very high, the number of people tested should increase in order to obtain a sufficiently low rate. The test period should also be extended long enough to constitute an adequate period based on survey data from actual use of the medicine or implant.

When there are concerns about side effects with long-term dosages or implants, the test period is the most important consideration. Since, as expressed in Equation (5.4), the target for the cumulative side effect rate decreases according to the reimbursement amount (E), and increases along with the corporation's capability for handling accidents (*a*) and the unit price of the item (*P*), the framework of the test experiments should change.

Hypothetically, for an accelerated test, the assurance period would be extended by the multiplication of the test period and the AF. Accelerated tests for new medicines cannot, of course, be implemented on human beings, but may be possible with animal subjects. For example, ultraviolet rays are possible harmful influences on living things, and are particularly activated in regions of high temperatures, so animal experiments in the daytime in the Middle Eastern desert could constitute such an accelerated test. Such an experiment could reduce the test period and stimulate the development of new medicines before clinical trials on the human body.

Let us summarize. First, we have established that there are no test specifications that apply indiscriminately to all products. We have also reviewed the core concepts and shortcomings of customarily used test specifications. And we have demonstrated the ease of evaluating the adequacy of test specifications by simple calculations. Of course, customers understand these principles as well as suppliers. If failures or side effects occur widely, customers might well compel corporations to reveal their test methods and the results obtained before the item was released. Corporations need to understand these test principles properly and apply them responsibly to ensure a safe society.

5.10   Advantages of parametric ALT

Parametric ALT has many advantages. Practitioners can easily configure their tests with assumed parameters, and CEOs and reliability chiefs can review the adequacy of test specifications, confirming the parameters and other factors. The most important insight is that BX life or failure rate can be expressed with figures. This leads to scientific decision making. Furthermore, if the test period must be decreased for speedy item release into the market or reducing item development costs, understanding the underlying basis of testing will generate alternative ideas for reviewing and analyzing the severity of the test conditions.

If differences arise between the test results and field data, the cause will be either mistakes in stress selection and the ensuing life-stress model formulation, or a mismatch with the parameters of the life-stress model. The accuracy of test methods can be improved by reviewing and adjusting the test conditions based on surveyed data about the environmental and operational conditions under which the item operates. Therefore, collecting data on the kinds of stresses and parameters experienced by the item in the field, or the exponent of outside stress, activation energy, and wear-out intensity (shape parameter), is crucial for developing reliability testing for the next model. CEOs can then confirm whether the mathematical conditions are appropriate for the proposed targets and that the failure mechanisms are identified and understood.

Meanwhile, if widespread reliability issues arise, the related specialists can assemble information and establish detailed design standards for the problematic items. It may be safe to design items according to established standards, but it also could limit the creativity of the designers. Developing parametric ALT that confirms failure sites in a short period will be more beneficial in terms of cost reduction or performance increases than sticking to the standards. Note that, like the severe test method of the fly-back transformer (FBT),* testing with a high acceleration factor can greatly reduce the confirmation time and allow assessing various ideas that would provide a competitive advantage.

At this point, we can compare the test results with the planned targets quantitatively. If the targets are not reached, products can be improved based on the analysis of the failures found. Sometimes products may be tested and improved several times in order to attain the target. Such a case reiterates that it is quite inadequate to decide about product reliability as simply "pass" or "fail," and that releasing the product without further improvements could result in enormous casualties. CEOs must pay attention to the quantified data, not to all-or-nothing results. Especially with the fierce competition for cost reduction now under way and its connection with corporate profit, targets for reducing costs necessitate extensive planning and innovative new structures or materials. In this climate, CEOs have to realize that releasing products without parametric ALT can sometimes turn apparent success into disaster.

A few cases can illustrate this. Michael Porter, a professor at Harvard Business School in the United States, wrote in the *Harvard Business Review* that the future of Japanese corporations was not optimistic because the operational effectiveness of Japanese corporations—their chief merit— was not the same as a basic strategy and therefore not a real differentiation point.[†] But Toyota pushed cost reduction, along with operational effectiveness, and was making greater profits than any of the three major carmakers in the United States.[‡] Cost reduction pertains not to operational effectiveness but to basic strategy, as long as it does not lead to problems. Toyota started implementing design changes for cost innovation on its compact cars in the middle of 1990. Then in 2000 it instituted a company-wide plan to evaluate, thoroughly and from the beginning, the tasks of

---

* See Chapter 5, Section 5.5.
† Michael Porter, "What Is Strategy?" *Harvard Business Review*, November–December 1996, p. 61.
‡ C. Park, "The Innovative Keywords through Global Innovative Corporation," *LG Business Insight*, May 3, 2006, p. 23.

four main agents—the developer, the designer, the production engineer, and the part vendor. This program, called CCC21, converted component improvement to system improvement, resulting in new components that integrated the functions of separated parts. This removed the waste inherent in interfaces.* Thus, Toyota could attain drastic cost reductions by applying new materials or new technology to system-level innovation. But problems related to reliability increased as well. Actually, the recall numbers rose drastically. As shown in the recall data, field reports of problems occurred at 6 months, 2 years, 3 years, 5 years, and 9 years after the first release of newly designed cars into the market.† It is hard to imagine that Toyota simply missed all the issues. More likely, some issues were uncovered, but satisfactory corrective actions were not taken, or those actions did not fully cover the 12-year expected automotive lifetime. Toyota's approval test specifications were clearly failing to estimate quantitative reliability indices and were therefore inadequate for identifying potential failure problems in newly designed products. It is essential that new products should be approved with newly established test specifications as much as possible in order to minimize problems and maximize profit.

Widespread reliability accidents will disappear with the general use of parametric ALT and a clear understanding by CEOs of reliability concepts and methodology. The pressure to push product development with reduced or eliminated test periods will diminish as CEOs recognize the need for clear reliability targets, execution processes, and quantified results. CEOs should promote using AFs and promise to support such efforts, rewarding good results. As CEOs increasingly understand the importance of reliability, they will not cut wide-ranging reliability operations, but rather maintain their functions and demand that their staffs study how to effectively improve their output. Their decisions would not be based on subjective policy judgments but on scientific conclusions. In the future, CEOs who do not understand reliability science will have a hard time taking charge of manufacturing businesses.

Now let's assume that we have obtained quantitative results with newly established test specifications and the results are far from the targets. Reviewing the failed samples has revealed the failure mode and the failure site before product release. Now it is time to fix the problem. The two pillars of reliability technology are identifying failures in advance and establishing corrective actions due to in-depth failure analysis, as mentioned in the prologue. So let's next consider failure analysis.

---

* S. Kwon, "Japan Corporation: Development Stage and Reinforcement of Cost Innovation," in *Automotive Review*, Seoul, Korea: Korea Automotive Research Institute, 2007-04, p. 36.
† The home page of Toyota recall: http://toyota.jp/recall.

Reliability integrity

chapter six

Failure analysis for accurate corrections

In order to make products highly reliable, three kinds of specialists must work together organically: the product developer, the reliability engineer, and the failure analyst. The task of the product developer, sometimes called the "know how" specialist, is to improve a product's performance and decrease its cost. The reliability engineer is the failure mechanism specialist who identifies reliability issues lying dormant in the product. The job of the failure analyst, or the "know why" specialist, is to analyze in depth the failures identified, both inside and outside the company, and to clarify their causes. The fields of these specialists are totally different. For a television set, for instance, the product developer is mainly an electronic engineer, the reliability engineer is usually a mechanical engineer with an understanding of failure mechanisms and parametric ALT, and the failure analyst is a materials engineer.

Let's consider the ways in which they work together. Product developers, who understand competition in the market, pass on information about the function and structure of the product, its reliability targets, and test samples to the reliability engineers. The reliability engineers complete tests and pass back to the product developers the two reliability indices (BX life and annual failure rate within lifetime) and estimated expenses for future reliability problems. In turn, they pass on data to the failure analysts about failure modes and failed samples identified in tests. Failure analysts, using these data and the design materials from the product developers, perform their analysis, report it to the reliability engineers, and propose alternative design changes to the product developer.

The CEO's task is to functionally differentiate the organization accordingly. First, the product approval section should be divided into two parts—performance approval and reliability approval. The section for failure analysis should be set up separately.

The performance approval area manages both the usual and specialized performance tests, but hands over the tests that lead to material strength changes to the reliability approval section. Reliability engineers identify problems that might occur in the future, executing reliability marginal tests and reliability quantitative tests. Thus, this section should be furnished with environmental testing equipment and the specialized test equipment required for item approval tests. The reliability section should be able to configure specialized test equipment according to test specifications, as well as to design parametric ALT.

The failure analysis section needs both frequently used nondestructive and destructive analysis equipment and tools for measuring material characteristics. Since it is impossible to purchase all the necessary equipment, failure analysts should identify nearby universities or research institutes that have the kinds of high-priced equipment they may occasionally need and establish cooperative arrangements for utilizing their facilities. The functions of such facilities are diversified due to advances in computer technology and measuring speeds. Using state-of-the-art equipment, they can nondestructively visualize material defects in failed test samples, and with a focused ion beam, simultaneously remove a thin layer of material for further analysis.

Failure analysts should be sufficiently knowledgeable to consult with specialists outside the company in order to increase the accuracy of the failure analysis. To do this, they need to understand all the processes of failure analysis in order to work with these consultants in reviewing the characteristics, limits, and operation of various types of analysis equipment. Sometimes, manufacturers are asked to analyze the failure of problematic items, which is like setting the wolf to guard the sheep. Since the item manufacturer has a high level of technical knowledge about the item, its failure reports are generally accepted, but it would be better to understand these as simply useful information, because the manufacturer is not likely to report results that might impact its company negatively.* Thus, it is important for CEOs to establish their own failure analysis sections and control them internally, apart from the input of any outside entity. To be trustworthy, failure analysis must be handled independently.

Mistakes in reliability guarantees can create serious problems with new products and lead to ensuing claims from customers, with corresponding increases in service expenditures after product release. If failures are not analyzed in depth and correctly, chronic problems could remain in the manufacturing processes, and production expenditures will increase as inspections and other temporary fixes are implemented.

---

* As American ethicist Reinhold Niebuhr wrote, in every human group there is more unrestrained egoism than the individuals who compose the group reveal in their personal relationships. Individuals may be moral, but it is difficult, if not impossible, for groups of people to view a social situation with a fair measure of objectivity. This is true of corporations as well. R. Niebuhr, *Moral Man and Immoral Society*, Louisville, KY: Westminster John Knox Press, 2001, pp. xxv, xxvi.

Once the reliability section identifies all the possible problems with a new product using parametric ALT and indicative failure samples, it is time for the failure analyst to take charge.

6.1   The significance of failure analysis

As we have noted, the media have widely reported accidents related to reliability in products made by advanced corporations in Japan and the United States. As described in Chapter 4, they can hardly avoid such incidents, given current reliability processes. But these problems can be prevented from recurring by executing thorough failure analysis. Companies making products of lesser quality do not solve these problems properly, although they occur repeatedly. In such companies, problem analyses are superficial and may be performed multiple times before the causes are accurately identified. Then, as the causes are noted, corrective actions are taken. Nonetheless, the problems often are not fixed, and companies experience chronic quality issues in their manufacturing and product development processes. Quality issues remain despite several efforts to fix them, and therefore quality defect rates do not decrease.

Why does this happen? These problems arise when the relationship between cause and effect is not completely understood. Corrective actions based on this misunderstanding will, of course, not be effective. If, however, the causes are scrutinized scientifically, then appropriate corrections will solve the problem. For this to happen, the analyst must be able to understand the structure, materials, and manufacturing processes of the item as well as possible received stresses due to environmental and operational conditions; extract potential failure mechanisms; grasp the functions of analysis equipment and consequently plan adequate analysis procedures; perform analysis according to plan; interpret and synthesize the results; and propose cost-effective corrections. If the analyst is doing the right things, the company's reputation for having failure issues will finally disappear as proper corrective measures are implemented.

The way to break the chain of events that lead to loss of functionality is to clarify the failure mechanism(s) causing the failure. Consider, for example, the failure mechanism of electrochemical migration. In this phenomenon, events unfold as follows: moisture infiltrates the gap between the packaged polymer and the sealed case, or directly penetrates the two electrodes; surrounding contaminants are dissolved and ionized, forming dendrites, which eventually grow to bridge the two electrodes. Proper failure analysis investigates the answers to the following questions: What caused the initial gap or crack? What is the source of the moisture? What is the path of the penetrating moisture? Which ions have been generated by which contaminants? How high is the voltage powering the electrodes? What material constitutes the dendrite? Necessary correctives then address removing the gap or the moisture, and so on, breaking the chain of failure events.

To build in higher reliability, manufacturers need to know as much about how things fail as they know about how things work.[*] To cost-effectively identify corrective alternatives, failure analysis should be conducted scientifically, in depth, and multilaterally. *Multilateral* implies that analytical conclusions can be reached through several different routes, confirming each other and providing high confidence in the conclusion. *In depth* means that basic causes must be clarified by pursuing successive issues to their roots. Failure analysis cannot involve guesses or assumptions, and it cannot resolve problems by simply replacing components or units identified as flawed.[†] Micromechanisms of failure must be researched to the lowest possible point—even the molecular level—by using various advanced measuring systems, such as scanning electron microscopes and analyzing the monitored images captured with them. This is the only way to interrupt the chain of failure-producing events and make the most effective corrections. Failure analysis must become truly analytic.

Let's consider the need for in-depth analysis from another angle. Every depth stage of technical analysis on a problematic issue has associated managerial measures that respond to it. The deeper the technical analysis, the more the basic cause is revealed, and the less managerial action is needed, resulting in lower expenses. Therefore, analysis should be focused on a microunderstanding of materials as well as on a broader understanding of item structures. Table 6.1 presents the problem of a malfunction in the review mode of a videocassette recorder, analyzed to six stages of depth. It is evident that the work required to respond to the sixth stage is less than that required at the fifth stage.

Advances in failure analysis have been accomplished in the last few decades largely through the contributions of materials engineers. Failure analysts used to think, like materials engineers, that the causes of failure are generally binary, allowing only two choices—either the part was defective or it was abused.[‡] But the failure analyst should check for three, not two, possible categories of causes: one involves abuse (whether users maintain and operate the product according to the user manual, and within design constraints), and two causes are related to defects (whether the parts are manufactured within the specified tolerances and the process capability is sufficient, and whether the related specifications are thoroughly and scientifically established). The three causes of trouble,

---

[*] M. Pecht, "Physics of Failure—Focus on Electronics," CALCE Short Course, University of Maryland, 2001, p. 5.
[†] W. Becker, ed., *ASM Handbook*, Vol. 11, *Failure Analysis and Prevention*, Materials Park, OH: ASM International, 2002, p. 319.
[‡] Ibid., p. 315.

Table 6.1 Managerial and Technical Measure of Analysis Depth

Depth	Cause analysis	Managerial measure	Technical measure 1	Technical measure 2
—	Not rotating when reviewed		Rotate properly	Rotate properly
Stage 1	Rocking of idle wheel	Inspection of rocking	—	—
Stage 2	Wheel moving up and down	Inspection of moving	—	—
Stage 3	Longer shaft length	Inspection of length	—	—
Stage 4	Injection condition change of shaft	Patrol check of injection condition	—	—
Stage 5	Worn-out injection mold	Periodic check of mold	New mold ordered	—
Stage 6	Mold made of low-grade material	Mold check when ordered	—	New material mold ordered

then, are misuse by customers, nonconformance with manufacturing specifications, and mistakes or superficiality in establishing specifications.* Of these, the adequacy of specifications is the essential reliability issue, which is frequently revealed only after analytic failure analysis. Confirming failure mechanisms by synthesizing the data from analyzing materials and stresses leads to accurate measures for establishing detailed specifications or for making material/structural changes. At this stage, several alternatives should be prepared for the product developer.

Advanced corporations fully understand the technology of failure analysis and apply a high level of systematic and scientific analysis technology to their products. CEOs whose products have chronic technical problems should recognize the necessity of implementing failure analysis technology. If chronic problems do not clear up with three or four iterations of internal analysis, CEOs should consider whether specialists from outside the company should be consulted. The in-depth experience of analysts is critical to solving problems precisely, which is why companies that specialize in failure analysis have survived.

6.2 Considerations in performing failure analysis

Let's think about the elements to be considered in failure analysis. Failure analysis is defined as a process that is performed in order to determine

---

* See Chapter 2, Section 2.4.

the causes or factors that have led to an undesired loss of functionality.[*] Failure analysis is an iterative and creative process, much like the design process, but with reversed roles of synthesis and analysis. *Creative* means that, in the absence of established, generally prescribed processes, practitioners must be innovative and broad-minded.

As with any other valid approach, the objectives of failure analysis should be clearly defined.[†] The analysis plan, including the range, procedures, and duration of the analysis, will differ according to its objectives: clearing failure in new product development, improving a product already in the market, assigning responsibility for failure, or simply preventing recurrence. The range of analysis should be set after consulting with relevant persons; securing and analyzing physical evidence should be performed thereafter. In finished reports, the investigative methods, results, and related data should be arranged systematically to ensure confidence in the conclusions and to ease the implementation of corrective alternatives.

One consideration in failure analysis is the principle called the *conformation bias*. This refers to the tendency to look only for what one expects to find.[‡] It is important for failure analysts to recognize the limits of their own knowledge, and to seek opinions from others knowledgeable about the product. Planning failure analysis, then, should be based on synthesizing various activities. These general preparations when planning include the following:

- Studying the functions, structures, constituent materials, and manufacturing processes of the subject item
- Investigating the received stress history due to environmental and operational conditions
- Consulting with failure mechanism specialists and analytical instrument experts
- Conducting interviews with supportive people who want to know what went wrong and with defensive people who may be concerned about being blamed for the failure[§]
- Selecting the sites to be investigated to highlight a particular type of failure

The most important thing in failure analysis is preserving the evidence. As we see in the media, nothing is more important in the investigation of fires or deaths than the preservation of the scene of the

---

[*] W. Becker, ed., *ASM Handbook*, Vol. 11, *Failure Analysis and Prevention*, Materials Park, OH: ASM International, 2002, pp. 315–323.

[†] Ibid., p. 317.

[‡] Ibid.

[§] Ibid., p. 321.

accident, because important information may not be immediately obvious. Likewise, failed samples may hide critical information.

For example, assume that the analyst observes inner and outer scars on bearings while trying to clarify the causes of a rotational noise. Before anything else, the bearings should be washed for examination by microscope. After taking actual images through the microscope, the analyst should examine the washing fluid for the contaminants that triggered the scars. If the fluid has been thrown away, there has been a failure to preserve evidence. To preserve and secure evidence, an analysis plan must proceed from nondestructive tests to destructive tests. But even placing samples under a microscope to observe the surface, a nondestructive test, means that something has been removed from the surface to obtain a clear picture. Completely nondestructive tests of failed samples are not really possible, and the best we can do is conduct *less* destructive tests. Still, it is paramount to preserve evidence insofar as possible. Furthermore, if good items as well as failed items are analyzed, analysts can easily reach conclusions by comparing the two to reveal the differences.

Failure analysts should thoroughly understand all kinds of failure mechanisms and analytic instruments. Because mechanical parts generally have a long history of continual advancement, their failure mechanisms are well established.[*] Since electronic parts comprise more and more varied materials and integrated microstructures than mechanical parts, their failure mechanisms are more diverse.[†] Furthermore, because electronic devices are continually progressing, analysts should be careful about identifying and announcing new failure mechanisms. Analytic instruments have also advanced rapidly due to new technology, so analysts should also become familiar with the functions and methods of new instruments and software through related conferences and exhibitions, and be able to apply them to their analyses.

In failure analysis, conclusions should finally be drawn by synthesizing analyzed results. Before making conclusions, it is important to check for judgment errors, such as ignoring a supposedly minor detail that could actually be significant, and to question once again any suspicious measurement data.[‡]

Finally, analysts should be careful to avoid logical leaps, especially those too easily made under the pressure of time. Even without such logical jumps, there is a problem with using the hypothetical-deductive

---

[*] W. Becker, ed., *ASM Handbook*, Vol. 11, *Failure Analysis and Prevention*, Materials Park, OH: ASM International, 2002, p. 557–1043.

[†] M. Minges, *Electronic Materials Handbook*, Vol. 1, *Packaging*, Materials Park, OH: ASM International, 1989, pp. 958–1057.

[‡] W. Becker, *ASM Handbook*, Vol. 11, *Failure Analysis and Prevention*, Materials Park, OH: ASM International, 2002, p. 319.

method to confirm scientific hypotheses—that is, the fallacy of affirming the consequent, as follows:[*]

1. If the hypothesis is true, then the prediction is true.
2. The prediction is true.
3. Therefore, the hypothesis is true.

This is an instance of the same false logic as "man is an animal; animals exist; therefore man exists." For the hypothetical-deductive method to be inductively correct in the confirmation of scientific hypotheses, it must include step 3, as follows:

1. If the hypothesis is true, then the observational prediction is true.
2. The observational prediction is true.
3. No other hypothesis is strongly confirmed by the truth of this observational prediction; that is, other hypotheses for which the same observational prediction is a conforming instance have lower prior probabilities.
4. Therefore, the hypothesis is true.

In other words, this hypothesis, among others, has a nonnegligible prior probability, or seems to be more probable.

This logical analysis reconfirms the importance of analysts remaining open to the opinions and advice of others in conducting a failure analysis, given the natural tendency to draw false but seemingly logical conclusions. Historically, there have been many cases when even great scientists have had to reverse their hypotheses. Einstein, who had been filling in the gaps in Newtonian physics, more or less reflexively inserted into his universe equations something called the *cosmological constant*, which he later called "the biggest blunder of my life."[†]

Dominant failures are identified with parametric ALT, newly designed before product release. Correctives established through in-depth multilateral scientific investigation are implemented for the product, and their effectiveness is confirmed by parametric ALT—indicating that all failures are cleared. The outcomes of reliability marginal tests, parametric ALT, and customer use tests should be categorized and arranged systematically. Applying these data proactively to the reliability engineering of the next model will decrease development expenses and time to market.

---

[*] W. Salmon, *Logic*, Englewood Cliffs, NJ: Prentice-Hall, 1984, pp. 132–137.
[†] B. Bryson, *A Short History of Nearly Everything*, New York: Broadway Books, 2003, p. 127.

6.3 Results after completing parametric ALT and failure analysis

Let's briefly summarize the process we have been describing. Hardware deteriorates physically over time and finally reaches failure. New products reviewed through proactive reliability engineering are tested via reliability marginal tests, and parametric ALT confirms whether the product's lifetime exceeds the target. When no failure is found, the accuracy of parametric ALT can be substantially elevated by extending the test period or adjusting the parameters of the test conditions. Multilateral and in-depth failure analysis on identified as well as chronic failures determines the failure mechanisms and suggests appropriate correctives, which are further confirmed as appropriate through parametric ALT. Finally, a perfect product is generated.

Repeating this process develops a design specialist of parametric ALT, the kind of reliability engineer described in the beginning of Chapter 6. Accumulating experience over time allows the designer to make increasingly accurate estimations of the types and locations of failures, and the value of parametric ALT is accordingly increased. Consider, for instance, one reliability engineer in charge of household refrigerators in a firm where I worked. With experience, his success rate at finding failures increased dramatically, from 30% to around 60%. As a reliability engineer's understanding of the results of failure analyses deepens, the specialist can identify problems with only a review of the drawings and propose alternative corrective actions. He can also make impromptu suggestions for expense reductions on costly structures and confirm his ideas with parametric ALT procedures he has designed. A perfect and competitive product results.

When I worked for company S, the product quality in a certain division had not been significantly improved for over 30 years in spite of various attempts to advance it. After a parametric ALT design specialist was developed, the product quality rapidly improved through his reliability validation. This was enhanced greatly by a company standard change in the product development process. Validation by parametric ALT and achieving a quantitative confirmation of reliability, in addition to conducting existing pass/fail reliability marginal tests, had recently been incorporated into the product development process. Although, of course, there were other excellent changes in aesthetic design and constant efforts to improve product performance, the success of the quality improvement program could be ascribed to developing designs using newly established test specifications for product lifetime. With them, the annual failure rate decreased and product lifetime increased. At that time, I executed a BX life test on every item (see Table 5.1) that dramatically improved the failure rate of core items, such as compressors and ice maker units.* After 1 year, the outcome was already apparent. In a study of major home appliances in 2005 by J.D. Power and Associates, company S's refrigerator scored 817 points on a customer satisfaction index, based on a 1,000-point scale, ranking first—22 points higher than the next competitor, and 45 points more than the third-ranking company. Moreover, the side-by-side refrigerator received greater public favor than the highest-ranking dishwashers and ovens, scoring 50 points more in customer satisfaction.

In our discussion so far, we have established that parametric ALT, together with scientific failure analysis, is the best tool for achieving perfect, successful products in the design stage. Although we have been trekking through the reliability jungle and overcome several difficult barriers, we have still not completed the journey. There remain notable pitfalls, as discussed in Chapter 7. If we relax our efforts in managing reliability and manufacturing quality, low-quality products will flood the market.

---

* J. Lim, et al., "Reliability Design of the Newly Designed Reciprocating Compressor for Refrigerator," *Proceedings of the 36th Reliability and Maintainability Symposium*, Union of Japanese Scientists and Engineers, 2006, pp. 179–183. J. Lim, et al., "Reliability Design of Helix Upper Dispenser in Side-by-Side Refrigerator," *Proceedings of the 36th Reliability and Maintainability Symposium*, Union of Japanese Scientists and Engineers, 2006, pp. 173–177. S.-W. Woo and M. Pecht, "Failure Analysis and Redesign of a Helix Upper Dispenser," *Engineering Failure Analysis*, 15, 2008, pp. 642–653.

chapter seven

Problems in the manufacturing inspection system and their solutions

The factors that managers generally consider most important are production quantity and production speed. This does not mean that other tasks are unimportant, but attaining the quantity target generally tops the list of a manager's priorities. However, if CEOs are overly concerned about production figures, quality will begin to decline. There is a trade-off between the quantity and the quality of products. If line inspections are skipped or out-of-tolerance items slip through, production quantity could of course increase, but the price would be high. It is particularly easy for production managers to overlook the kind of quality problems that are not immediately noticeable. To what tasks in the manufacturing line should CEOs pay particular attention? The answer lies in the slogan encountered frequently on construction sites: safety first, quality second, and production (quantity) third.

Safety is obviously paramount. An accident that claims an employee's life is an almost indescribable tragedy. It would cause the morale of workers to drop and, without any countermeasures, cause distrust of management. Accidents related to safety are generated by ignorance and inattention.[*] Moving from ignorance to safety can be achieved through consultation with specialists, taking corrective actions, and instituting periodic training.

Production problems—malfunctions in the facility, a shortage of vendors' materials, and so on—delay product shipment. But such disturbances do not usually grow into major problems. Although they seem critical at the time, they are often like a tempest in a teacup. When they occur, the production section looks bad, and lost opportunities will mean the market share for the product will decrease a little. But this does not mean that the company will lose its market share. If the product itself is trouble-free and has comparative advantages over its competitors, the losses due to production delays can be easily recovered. Moreover, solutions to production problems, such as enhanced production methods and productivity improvements, are

---

[*] D. Ryu, *Corporation Revolution* (in Korean), Paju, Korea: Hanseung, 1997, p. 91.

already established and easily available through specialists and training institutes dealing with production systems. When a production problem occurs, it is best if the CEO expresses interest and observes the staff's handling of the problem, being careful not to intervene directly. Otherwise, the entire organization becomes focused on production above all other functions, while safety and quality issues continue to grow.

On the other hand, quality accidents can have a disastrous aftermath, as we have discussed. Therefore, they must be addressed before entering production. Good quality materials make it easy to produce good final products, and products planned for high quality are easy to sell in the marketplace. Thus, CEOs should focus their efforts on improving product quality once a product development plan is determined. Especially if the planned product is completely new, the CEO should concentrate on raising its quality to the first rank and keeping it there, regardless of possible product release delays. Over 20 years ago, company S in Japan decided to switch to VHS videocassette recorders from the Betamax type, which was the company's original format, in order to beat the competition. The launch time of the new item was delayed a year later than executives expected—a necessary trade-off allowing company S to match the quality of the new product to the company's reputation for first-class products.

If CEOs address internal issues that will improve product quality, production and sales will take care of themselves. If they look deeply into the circumstances under which managers try to solve quality problems, CEOs will understand that measures to improve product quality are relevant to all aspects of the company and need to be managed from the highest level. Staff managers are often not in a position to resolve quality issues on their own. For example, it would be hard for the quality chief to independently add processes requiring further investment, or to supplement his section with staff from other areas of the organization. Once such overarching activities are addressed, practitioners can generally produce and sell products well enough. So, in companies whose products demonstrate inferior quality, it may be that the CEOs do not thoroughly understand the quality of their products and how to achieve it, even though they may think they do.

First, CEOs must recognize the vulnerability of the quality inspection system. If they cannot enumerate how many quality issues their products have, or there are more than 10 such problems, they are likely to have further problems. When the number of quality issues reaches double digits, it is difficult to keep track of and properly manage them, especially since all the relevant personnel must understand and respond to them. In this case, it is better to stop production as soon as possible and establish an overall plan for improving quality and implementing planned correctives to both production and staffing. Until such time as the product achieves the level of quality of similar products from leading companies, improvement activities should be carefully monitored and controlled. As the final gatekeepers who are closest to the eventual customer, manufacturing managers should find and fix any problems that the product developers and reliability engineering staff have missed. If it is impossible to realize a degree of quality equivalent to that of the leading producers in a short period—say, 1 year—CEOs should accept the limits of their firms' technology and consider outsourcing quality technology and management.

7.1 The architecture of the manufacturing inspection system

Crucial to improving quality are the architecture of the inspection system, methods for quality stabilization of manufacturing lines, and considerations of design changes. The quality inspection system is often inadequate.

Manufacturing inspection systems are generally classified in order of processing: incoming inspection (II), process test (PT), final test (FT), and outgoing inspection (OI). Dividing these phases according to inspection characteristics identifies two types: total inspections and sampling inspections. Process tests and final tests in the line are total inspections; incoming and outgoing inspections are sampling inspections.

First, let's discuss how to calculate whether or not, in terms of inspection power, inspection systems are designed properly. The power of a total inspection is defined as the ratio of the number of inspection items actually in process to the number of items to be inspected:

The power of total inspection = (The number of items in process

[PT + FT])/(The number of items to be inspected)   (7.1)

where the items to be inspected are determined by customer order specifications, such as electrical power (say, 220 V, 60 Hz) and performance requirements, and by projected future problems. It would be good, but impossible, to inspect all the checklisted items for each product according to the given specifications, but many items with a low probability of being problematic are inevitably omitted in a total inspection for the sake of manufacturing efficiency. Thus, the power of total inspection is usually about 50%, differing from product to product and among technological levels.

It is very dangerous to omit any questionable inspection items, so sampling inspections are conducted. But the smaller the sampling ratio and the longer the sampling period, the greater is the risk that there will be defective products. Using sampling inspections implies a willingness to take the loss if any problems occur because of insufficient inspection.

Acknowledging this, let's calculate the power of inspection, including sampling inspections as well as total inspections:

The power of inspection = (The number of inspection items in process

[II + PT + FT + OI])/(The number of items to be inspected)   (7.2)

This value cannot reach 100%. To approach 100%, a patrol check (PC) should be introduced. Patrol checking means that inspectors are overseeing the process line and inspecting products in process. Because some damaged products are inevitably flowing through the assembly process, it is a best practice to check products inside the company for some issues. Sometimes an equivalent inspection can be conducted by dismantling a product after outgoing inspection, but it is difficult to increase the sample size and, since the product may be damaged during dismantling and reassembly, the process itself introduces flaws.

The CEO expects the manufacturing chief to closely manage his area, frequently checking the workmanship of the product. But the head of manufacturing also has many other responsibilities. He must frequently go off-site to participate in conferences and meetings with vendors, so both he and the CEO might feel uneasy about quality control. It would relieve this unease to have a patrol checker on the job. In this case, Equation (7.2) changes as follows:

The power of inspection = (The number of inspection items in process

[II + PT + PC + FT + OI])/(The number of items to be inspected) (7.3)

The goal should be to approach 100% as closely as possible.

In performing a patrol check, it is best if the number of items to be checked is determined by how much one or two checkers can handle, adjusting the number of items and the checkers' work shifts as necessary (this will differ from site to site) in order to raise the power of inspection as much as possible. Critical-to-quality inspection items that might possibly cause failure in the market should be selected as patrol check items, with the results recorded and preserved in patrol data sheets to aid in estimating the possible aftermath of any failures. The patrol checker should use his own instruments insofar as possible to assess the accuracy of line measurements, to ensure that the instruments are accurate and properly set. In addition, some inspection items should be added to check the level of the operator's skill and the accuracy of the facilities, both to establish the accuracy of the equipment and to guard against the poor implementation of work standards or the misalignment of setup tools.

We have stressed that the inspection system cannot be perfect. Moreover, if the power of inspection is raised, manufacturing costs increase, while if the power of inspection is lowered, after-sales service expenditures increase. This is a dilemma for management. Can it be resolved? Let's start to answer that by reviewing the adequacy of the inspection system quantitatively. A high level of quality means that damaged or defective materials and components are eliminated before manufacturing begins, and that products with defects due to poor workmanship or other causes are not released into the market.

Let's consider the relationship between the quality level of the product and the quality defect rate screened out by inspection processes. This is called the in-house defect rate. The fact that the in-house defect rate is low does not mean that the product has a high level of quality; the low level of the *in-house defect rate* can also be obtained with improper screening. High quality means that the market defect rate is as low as the in-house defect rate. Conversely, the fact that the in-house defect rate is high does not mean that the product is of high quality; remember that the power of inspection cannot reach 100%, since not all items can be inspected. However, we can fairly assume that the defect rate of the screened product would be the same for unscreened items, when the power of inspection is 50%. For example, under the above power of inspection, if the in-house defect rate found through the inspection system is 2%, there should be a 2% defect rate in the market. This defect rate, called the *undetectable rate*, is estimated as follows:

$$\text{Undetectable rate} \cong \text{In-house defect rate}$$

$$(1/\text{The power of total inspection rate} - 1) \tag{7.4}$$

Using this, let's interpret the quality level at the manufacturing site. Since certain defective components, especially those related to failure, are difficult to find through inspection in the manufacturing line, the market defect rate will be higher than the undetectable rate. Thus, there are three possible situations, as follows. Situation III is the goal.

Situation I: Undetectable rate ≤ market defect rate target ≤ market defect rate.

Situation II: Market defect rate target ≤ undetectable rate ≤ market defect rate.

Situation III: Undetectable rate ≤ market defect rate ≤ market defect rate target.

In Situation I, the inspection activities at the manufacturing site are dubious. While the undetectable rate should be smaller than the market defect rate target, it is unsatisfactory if the actual defect rate in the market is the highest of the three. Proper inspection inside the company will bring the undetectable rate closer to the market defect rate, though it cannot exceed it. As shown in the first line of Table 7.1, a market defect rate of 4.5% is much greater than an undetectable rate of 1.0%. This large difference between the undetectable rate and the market defect rate indicates that managing the inspection process for this product is a mere formality. Perhaps the manufacturing head in this situation insists that quality issues are due to design mistakes or that it is not possible to adhere to the specifications of a difficult design. While of course it is essential to identify any design problems once the item goes to the manufacturing site, it should nonetheless be required that manufacturing chiefs reach an acceptable in-house defect rate target, which can be calculated as follows:

$$\text{In-house defect rate target} \cong \text{Market defect rate/}$$

$$(1/\text{The power of total inspection rate} - 1) \tag{7.5}$$

In order to achieve the targeted in-house defect rate, the inspection activities should be reviewed. Adequate inspection standards, accurate measuring instruments, and sufficient inspectors should be provided; if there is no recent paper trail of revisions of inspection standards, improperly adjusted or malfunctioning measuring instruments, blanks in data sheets, and inspectors who frequently leave their positions, the power of inspection is only a meaningless figure. CEOs should require their staffs to develop a robust inspection system and train inspectors to do their jobs according to current inspection standards.

The result of proper inspection, after these actions are taken, is shown in the second line of Table 7.1. This constitutes Situation II. Situation II obtains when the undetectable rate is greater than the market defect rate target, which occurs when inspections are done properly. As shown in the third line of Table 7.1, after a review of the inspection system, additional inspection items should be added to raise the power of inspection. The sum of the in-house defect rate and the market defect rate on the second and third lines of Table 7.1 (5.5%) should be screened inside the company. When the market defect rate target is 1%, the power of inspection should be raised to 0.82 (= (5.5% − 1%)/5.5%). But something else must be done before the power of inspection can be raised, since raising it also escalates inspection expenses. This is an audit of the inspection system by manufacturing engineers and product developers. In addition, products must also be inspected and tested thoroughly using customers' data, starting with reports of problems encountered in use and proceeding to previous processes in order to progressively inspect from products to units to components. Any loopholes in the inspection system should be identified. For auditing, the sample size is increased based on the defect rate and inspection items are added according to comparisons of the distributions of parts to their tolerances. If the inspection system is supplemented, the power of inspection rises and the in-house defect rate also increases.

Table 7.1 Analysis of the Inspection System[a]

	Power of inspection	In-house defect rate	Market defect rate	Undetectable rate	In-house defect rate target	Issues	Activity
1	0.5	1.0	4.5	1.0	4.5	Improper inspection	Training inspectors
2	0.5	2.0	3.5	2.0	3.5	Target not attained	Increase the power of inspection
3	0.6	3.0	2.5	2.0	3.75	Many defects	Improve at origin
4	0.6	1.8	1.7	1.2	2.55	Not attainable by inspection	Aging test
5	0.6	1.8	1.2	1.2	1.8	Near to target	Increase the power of inspection
6	0.7	2.1	0.9	0.9	2.1	Target	—

Note: The market defect rate target is 1.0%; the market defect rate of a competitor is 1.2%.

[a] The ratio of the in-house defect rate to the undetectable rate is equivalent to the ratio of the power of inspection to (1 – the power of inspection). The greater the sum of the in-house defect rate and the market defect rate, the greater is the number of tasks to be handled.

When the in-house defect rate approaches the target set for it, many defective items are being found in the line. This is the point at which efforts to eradicate quality defects should be implemented by applying the internal technology intrinsic to the products to solving them. The more defective items are identified, the easier it is to solve problems. The activities of the inspection system should also be simplified, as shown in the fourth line of Table 7.1. In the case represented by Table 7.1, the sum of the in-house defect rate and the market defect rate in the third line of Table 7.1 (5.5%) was decreased to 3.5%, as shown in the fourth line.

If quality issues are solved at their origin, the power of inspection should be recalculated because the number of items to be inspected decreases. When the in-house defect rate target is not attained, as shown in the fourth and third lines of Table 7.1, both the inspection system and the quality system should be revisited. The 0.5% difference between the market defect rate and the in-house undetectable rate (fourth and third lines of Table 7.1) cannot be reduced solely through the inspection system, but can be improved through some particular program identifying failure, such as an aging test or parametric ALT. Some problems that occur in the market cannot be addressed or are difficult to find through an inspection system, which is evident when they are investigated one by one. Issues of performance and reliability in special environments and long-term use and minor damage incurred in the manufacturing line are not easy to detect. These all are due to a failure to recognize design mistakes or to adequately pinpoint critical-to-quality parameters in manufacturing. Therefore, a quality system in the manufacturing section that investigates and identifies mistakes in the design and manufacturing specifications should be instituted. To deal with these problems, special quality system programs should be established temporarily, and the results should be carefully reviewed. Some programs may be maintained continuously throughout production. Sometimes managers need to change their thinking completely and assume a state of emergency, treating the product as though it is newly developed and flowing in a new production line. Product verification, the adequacy of sequencing processes, and operator training should be reviewed again.

When the in-house defect rate meets the target after reforming the quality system, ideally the undetectable rate inside the company will match the market defect rate, as shown in the fifth line of Table 7.1. Here, the sum of the market defect rate and the in-house defect rate is 3.0% and the market defect rate target is 1%; thus, if the power of inspection increases to 0.67 (= (3% − 1%)/3%), the goal would be achieved. As the number of inspected items decreases as problems are solved at their origin, it is relatively easy to increase the power of inspection.

The sixth line of Table 7.1 pertains to Situation III, or planned target attainment. Here the sum of the market defect rate and the in-house defect rate, or 3.0%, reveals the troubleshooting capability of engineers inside the company. The remaining task is to move this sum toward 0%. CEOs should review whether this can be accomplished using their own technology or by using outside specialists. In any case, a defect will not disappear by itself.

CEOs aiming at first-class companies require their staffs to establish and report the power of inspection, in-house defect rate, market defect rate, undetectable rate, in-house defect rate target, and market defect rate target. They should review how much the defect rate decreases, no matter how the target is reached. If CEOs required subordinates to explain quality with this kind of quantitative data, personnel would recognize how vulnerable the line inspection system is and be able to master one of the most basic issues in manufacturing management. Let's follow the sequence as shown in Table 7.1.

While product development is analogous to attacking during a war, product quality control is like establishing wartime defenses. Omitting some of the requisite inspection items or neglecting to educate inspectors would be to neglect defense activities. What's more, inspections and screenings are not all of the necessary tasks; organizations must also work to simplify the inspection system and achieve a target of zero defects inside the company.

It should now be clear that inspection systems are imperfect. Some issues related to product quality that are missed in the product approval process would be difficult to find through the inspection system. Therefore, the quality system—a broader concept than the inspection system—must be addressed systemically. The quality program includes various systems, such as product approval systems (reliability marginal tests and parametric accelerated life test), systems that manage operator expertise, and systems that check facilities precision, as well as inspection systems. These are discussed in the next section.

7.2   Quality improvement and stabilization of manufacturing lines

A newly designed product begins production on a new line staffed by new employees in newly established facilities. At this opening stage, there are both doubts about the perfection of the product design and difficulties confirming whether the sequencing and configuration of processes are appropriate. The manager also must be assured that the training of the operators is sufficient. After many adjustments, the new product comes out and passes its final performance tests. Is it ready to release into the market?

Not necessarily. It is not easy to secure outstanding engineers to establish new quality systems, especially since product design and the setup of manufacturing lines also require good engineers. The design specifications may be confirmed with too little time before product release for the quality staff to test all the quality specifications related to the new product. No matter how quickly they are designed and implemented, they will be doubtful from the standpoint of accuracy. To mitigate these difficulties in one case, company S in Japan gives the director of the new product division the power to choose necessary personnel from throughout the company.

Quality systems for a newly established product should be different from those used for stable lines of existing products. Even for stable products, if quality innovation is necessary, dimensionally different activities need to be implemented. Innovation starts with diagnosis, the results of which become the basic material of planning. In this case, diagnosis means the investigation of the product as the outcome of all the activities that constitute its development and manufacture. How well both quality systems and products are devised and executed can be judged from the state of the final product. What should be investigated are problems not only with the product, but also with the manufacturing line and the workforce. Additionally, attention must be paid to such issues as poor response time and negative attitudes from product developers and engineers.

How can the CEO recognize and address all these problems? The only way is to discover, through testing in advance, issues that might weaken the final product and its components. These issues can be solved by the staff, but the CEO must convince quality-validating personnel of the importance of increasing the number of tests. While the responsibility for a problem occurring in the market can be argued and create staff conflicts, problems identified inside the company can be easily solved with full cooperation. The more carefully the causes of problems are clarified and identified internally, the more appropriately and rapidly quality staff can consult with the responsible areas and the greater will be the effort of all concerned. Therefore, it is good practice to produce only a few products at a time and to test them all. Finding problems through internal testing is the least expensive alternative; likewise, identifying trouble through domestic customer use is cheaper than hearing about it from overseas customers.

In a newly installed line, several steps should be taken to achieve thorough testing. First, assuming various specialized environmental and operational conditions, more test methods should be applied to the product than those used by design approval. Second, all products should be age-tested while testing mass production (TMP), because an increased test sample size makes it easy to find design flaws or minor manufacturing damages that are not revealed through product approval tests. Third, life test products should be sampled from mass production (MP). These tests can investigate and confirm whether design verification has been properly executed, whether there are any differences in components and manufacturing methods between the prototypes and the mass product, and whether the approval data for a similar model already in production can be substituted for the design approval data on the basic model without a complete round of new approval tests.

At one point, I developed a household air conditioner that was on the verge of going into mass production. A newly organized group of product developers had designed the product and a newly assembled workforce produced the air conditioner and its core unit, a rotary compressor, with all new facilities. I asked the managers to develop and configure one new test method a week to add to the generally acknowledged test methods. About over 100 kinds of tests were developed on the compressors and air conditioners and applied to find problems, which were then fixed in advance of product release. Consequently, there were no claims or expressions of dissatisfaction after overseas sales.

When I took charge of producing videocassette recorders, the annual after-sales service rate for existing models was over 10%. This meant, realistically, that it was impossible to sell the recorders. Before mass production on a newly designed model, the production line and quality system executed tests of mass production (TMP) several dozen times. Because the product subsequently exhibited high quality, the annual after-sales service rate dropped to 0.7% in 3 months. About 50 recorders were assembled in the daytime and tested all night, using aging tests; the problems were fixed, and the recorders were again disassembled and reassembled the next day, and so on, for 3 weeks. The repetition of these operations brought about various improvements in chronic issues, such as identifying problems, confirming the quality system, training workers, and eliminating discord between the development and manufacturing areas. It took only 3 weeks to rid the product of problems. Knowing that an advanced company does not carry out aging tests may lead a smaller company to think there is no need to conduct them, but they must be performed until problems no longer occur. If aging tests do not appear to resolve all product problems, a new program, such as lifetime testing, should be introduced.

When I took charge of a business unit producing reciprocating compressors for refrigerators, I told the production manager, who wanted to post the daily production table to the office wall, not to do so. Since it was his job to manage production quantity, I asked him instead to post the daily defect rate and amount. As quality became highlighted as a factor hindering production quantity, I had the quality manager attach a chart showing the daily state of quality and prepare to test 300 compressors.* Two compressors were sampled in the final test each day, and six sets from three assembly lines were tested continuously for 50 days under severe conditions. As the testing of as many as 300 compressors was carried out, test vehicles were designed with simplified architectures and made less

* See Chapter 4, Section 4.3.

expensively. Thereafter, if failures occurred, managers in the appropriate area analyzed them, established corrective actions, and fixed them.

Finally, two issues remained that they could not fix: an electrical short circuit at the small radius of curvature in the enamel coil of the compressor motor and the early wear-out of the crankshaft due to out-of-tolerance machining by machine tools that were not being well maintained. The corrective actions for these issues were to introduce new machines into the motor production line to form the motor coil and improving the precision of the grinding machines that made the crankshaft through proper maintenance. As you might imagine, there was a deep conflict between the product developers and the manufacturing staff as the necessity of additional processes was suggested. The production unit insisted that the motor as designed was difficult to make. In the long run, the improved compressor became a first-class quality product and the company was flooded with orders, which allowed its price to rise by about 10%.

In-depth analyses of issues make the inspection process unnecessary.* For this to occur, the capability of the technology related to the quality issues must be strengthened. For motor problems, it is difficult to screen estimated defects in the enamel wire; for crankshaft troubles, inspections are possible but unnecessary. The addition of a new process and the adoption of corrective maintenance were addressed by the technology intrinsic to the issues. Since technical competence was high, the company could feel confident that the problems were solved.

Generally, since the printed circuit assembly in a videocassette recorder is a combination of components already successfully applied to other products, it would seem that its targeted lifetime was ensured. Nonetheless, its failure rate would be uncertain, due to the possibility of inadequate configuration in the new circuits. Such problems in an electronic apparatus are found mainly by failure rate tests. On the other hand, any minor change in a mechanical item, like a compressor, can trigger multiple failures and decrease the lifetime. Therefore, mechanical items must be reviewed in light of the lifetime as well as the failure rate. In addition, the possibility that a product like an air conditioner would be operated under unexpected conditions must be taken into account. In this case, failure rate testing (or aging testing for 3 weeks for a videocassette recorder), lifetime testing of 300 compressors for 50 days under severe conditions, and reliability marginal testing of air conditioners in over 100 different environmental conditions led to adequate specifications and a reliable product.

We have discussed several considerations regarding newly established products and their quality issues. Now let's discuss operating manufacturing lines. Simplifying the inspection system means installing inspection

---

* See Chapter 6, Section 6.1.

processes at as many earlier stages as possible, changing total inspections into sampling inspections, and increasing the interval between sampling inspections. In order to do this, core components and processes should be analyzed statistically. First, essential specifications should be reviewed and their adequacies identified, and processing capabilities (C_p) should be assessed with measured data. Since inspections can be omitted if the process capabilities are good enough, methods for improving them should be considered. Since it is impossible to do the checking if there are no adequate inspection instruments, practitioners in this situation become discouraged and, eventually, simply bystanders. A workman is only as good as his tools. Sometimes the CEO will need to determine whether the appropriate inspection instruments are available, both for accurate inspections and to encourage top performance from the quality staff. In addition, manufacturing engineers should execute reliability tests specific to the manufacturing site in order to identify omitted or inadequate specifications and to be able to respond to future changes in components or processes.

Reducing the inspection system requires, as we said, upgrading the process capability; this in turn means narrowing item distribution and raising the level of workmanship. Ultimately, the various quality controls at the manufacturing site converge on two points: precision control in the facilities and management of the workmanship. Managing according to the type of work operation is summarized in Table 7.2. Here, the "tool" refers to a working tool in an assembly line—dies and molds in a press and injection molding shop, and jigs and fixtures in a machining shop. The accuracy and precision of any tool, machine, or cutter will wear down over time. Therefore, a maintenance team should be organized to periodically confirm the precision of the tools and to repair them to increase precision. Moreover, since machining requires one more item to be managed than does pressing/injection molding, the latter process should be used as much as possible, which will improve quality and reduce cost.

The quality system should be determined by assessing the current technological capability of the company to solve problems itself; therefore, it is never adequate to simply adopt the systems used by other, first-class companies. If a given company has a low level of technological capability,

Table 7.2 Management According to the Type of Work Operation

Line/shop	Workmanship	Precision of facilities		
	Operator	Tool	Machine	Cutter
Assembly	√	√		
Press and injection molding	√	√	√	
Machine	√	√	√	√

its inspection processes must increase; otherwise, service expenditures will increase and the market share of the product will decrease. If a given company has a high level of technological capability, the quality system can be simplified. One very stable company I know carries out only two simple total inspections at one process spot and at one final stage before shipment, without any outgoing sampling inspection. Reliably eradicating problems minimizes the extent of the inspection system.

Confirmation processes consisting mainly of inspections in the early period of mass production do not need to be performed in the stable period afterward. Automatic facilities for total inspection of some items become unnecessary once flaws are permanently fixed. The investment in the effort of solving problems should keep moving forward, with decisions about the investment in related quality systems made afterward, depending on the development of processes for resolving issues. The investment in an inspection system established in response to earlier quality issues may sometimes turn out to be futile.

The ultimate task of the quality system is to monitor the precision of the facilities and operators' workmanship and to know if they decrease. If this is not done, the whole quality system will deteriorate a little day by day without being noticed. Operators should be trained to keep to the established system without fail. Finally, the system should be kept as simple as possible to keep expenses low.

7.3   Considerations regarding design changes

As we have discussed, it is not possible to achieve a perfect inspection system due to the inherent characteristics of the system itself, so all problems should be solved at the beginning with related technology. There is also another issue. Because reducing material costs and manufacturing expenses is an ultimate goal, changes are always occurring in a manufacturing company. All too often, the company fails to respond to these changes, which results in quality failures. The quality accident incurred by company S of Japan in 2004 with its charge-coupled device (CCD) indicates a failure to manage change.[*] In short, it missed the significance of changes in materials and process conditions.

In order to prevent such occurrences, a system should be established to routinely process information related to change. All change points should be subject to advance reports and consultation. The term *change point* references details of facilities or mold/die replacement, material exchange, altered manufacturing methods, replacement or training of

---

[*] N. Asagawa, "Were CCDs Screened to Prevent," *Nikkei Electronics* (Japanese), November 21, 2005, p. 97.

operators, and so on. It is a serious undertaking to manage all change points. Practitioners should review and address change points regularly. Sometimes section chiefs need to review practitioners' decisions and the results over longer periods. If it is necessary to conduct physical tests, changed parts or units should be tested regularly at the same time each month for approval. It is important to note that change points are also generated from both inside and outside the company. Frequently, only outside change points are handled, while those occurring inside are neglected. All must be managed.

Even with a system in place to check change points, a few points are often missed. If productivity improves, product quality should be checked for at least 1 month and compared to that before the change, because there is a trade-off between the quantity of production and the quality of the product. If productivity increases due to changes in working methods and facilities, the process should be examined in advance, as it involves major changes. Because changes in materials or structures for productivity improvement and cost reduction cause critical accidents due to changed failure mechanisms, not only quality defects but also the possibility of failure should be investigated. Therefore, middle- or long-term lifetime tests should be established in the manufacturing area, regardless of product approval test results.

Company S, as mentioned, missed this point. Material change always involves reliability issues, the confirmation of which would be impossible through existing test specifications.* Let's look at one established test method as an example. Let's say a new paint or precoating material for an automobile is developed. This necessitates coating vehicles and exposing them for 2 years under the blazing sun in the U.S. Arizona desert; the results become the basis of judgment for the use of the material afterwards. Likewise, all material changes must go through reliability testing or parametric ALT. Many structural changes should also go through parametric ALT, though the conditions vary.

Now let's consider the omission of declarations by subcontractors. Although subcontractors are required to report change points, they do not always reveal all minor changes. For example, they might not report an increase in their defect rate from 0.01% to 0.03% due to neglected or improper equipment maintenance, while advertising a 15% productivity increase. As another instance, subcontractors do not reveal the composition of rubber items to buyers, considering this secret information; they change the ingredients to achieve cost reductions and do not inform buyers of the details. Nor would they report something that would raise suspicions about their technological or managerial capability. Therefore, an

---

* See Chapter 4, Section 4.3.

additional article should be inserted into contracts with subcontractors, assuring that they will pay a predetermined charge for failing to report changes, and that they must provide reimbursement for service costs due to unreported changes.

From a different angle, an event such as a change in the quality staff of a subcontractor or the serious illness and absence of the company's president could potentially affect the quality of the subcontracted items. By the same token, a sudden influx of many orders from other companies could alter the quality of the subcontractor's products, including existing ones. An increase in material prices or a shortage of materials could influence quality as the supplier substitutes other materials, even if the new material has the same specifications as the original.

Keeping quality constant is not easy. Again, it is critical that manufacturing chiefs as well as CEOs pay attention to issues related to quality, while entrusting experienced practitioners with production itself. Executives should be thoroughly informed of all developments, inside and outside the firm, that might possibly affect quality. A careful analysis of the entire process of receiving information, classifying and synthesizing it, designing and executing responsive tasks, and managing the results will help ensure a good quality information system.

Yet another task must be addressed. Although establishing measures responding to signals from the market occurs after the fact, it must also be systemized. If peculiar troubles occur in the market, they should be considered as early warning signals and further accidents should be blocked at an early stage. Without such a system, all trouble reports could be regarded as referring to already identified problems. Company S may have made mistakes in this regard. It must have missed a signal indicating that the outdoor equipment that incorporated the CCD was used in Southeast Asia under more severe environmental conditions than those in Japan. If the failures had occurred within 18 months after the product was released into the Japanese market, they would have occurred in about 9 months in Southeast Asia; the degradation rate there was twice as fast as the Japanese rate, since the temperature of Southeast Asia is higher than Japan's by about 10°. Anyone who understands parametric ALT would agree. CEOs need to pay particular attention to the reliability data of subtropical areas, as they indicate what will happen later in temperate regions.

Finally, as far as polymers or composite materials are concerned, it is essential to receive change reports to avoid major reliability accidents. But it is not enough to simply ensure that they are reported. An analysis method must be established through which material changes are revealed rapidly, and test methods that can confirm reliability in a short period are developed quickly.

Thus far, we have addressed basic quality. If these steps are adopted, we need not worry about plunges in market share. Even better, if the product has advantages over its competitors' in important attributes that customers appreciate, it will become a hit and prevail in the market. Chapter 8 deals with the methodology of customer satisfaction and comparative advantage.

section four

Product integrity

chapter eight

Customer satisfaction and comparative advantage for new product success

In 1960, Theodore Levitt, a professor at Harvard University, published a chapter titled "Marketing Myopia," which emphasized the significance of customers.[*] Levitt preached the concept that the purpose of a business organization is not to produce or sell products but to buy customers. His ideas about customers have attracted broad attention and have been making CEOs face the fact that customers decide their destinies. This has led them to rethink their business visions. Now customer satisfaction is recognized as the indispensable ingredient to be considered within the corporation, and all business organizations exert multilateral efforts to elevate the degree of satisfaction. Although this has produced good results, it doesn't seem that their market shares have increased as much as they expected.

To explain this, let's consider the concept of competition. Existing competitors, as well as customers, influence the outcome in this age of red ocean competition. In 1982, K. Ohmae of McKinsey in Japan highlighted the existence of competitors as the strategic factor in the concept called the *three C's model*. The three C's model delineates a strategic triangle, the vertexes of which are the corporation, the customer, and the competition.[†] The relations among the three C's should be understood by anyone strategizing product success.

The market is a battlefield on which a given company and its competitors scramble to draw customers for a product. The market share, or the outcome of the battle for customers, will be decided by product value. The two elements of product value are price and market quality.[‡] If products are competing at the same or almost the same price, the three C's model helps evaluate a competitive edge in terms of quality, both for customers

---

[*] T. Levitt, *Marketing Imagination* (Korean, English title: *Marketing for Business Growth*, 1974), Paju, Korea: Twenty-First Century Books, 1994, p. 145.

[†] K. Ohmae, *The Mind of the Strategist: The Art of Japanese Business*, New York: McGraw-Hill, 1982, p. 91.

[‡] See Chapter 2, Section 2.1.

and against competitors. In order to prevail in the marketplace, a corporation should take these two qualitative high grounds: full customer satisfaction and better product performance than the competitor's.

Whether a product is fully satisfactory or not can be investigated and rated quantitatively through customer satisfaction surveys. With those data, the strategy for product development can be established scientifically. In order to develop activity plans, in-depth surveys on unsatisfactory product attributes should also be performed. The survey results will vary from attribute to attribute, so information will come from different sources, depending on the issue being investigated. For attributes related to after-sales service, customers who requested after-sales repairs and service representatives should be surveyed; for attributes related to performance or additional features, it is necessary to investigate all products, including those of competitors, comparatively, in terms of the competitive edge in performance fundamentals and features. No matter how well most attributes of a product are satisfied, if its core performance does not put it in an advantageous position over competitors, customers will turn away from it. Although performance fundamentals differ from product to product, their advantages become differentiation points. Since the more complex a product is the more diverse its performance fundamentals will be, a corporation should secure comparative advantages over its competitors by improving performance fundamentals.

In order to beat competitors in a globalized market, a corporation should produce products of equivalent quality at a lower price, with higher customer satisfaction in important attributes, and superior advantages in core performance. Let's examine quality in the broad sense, or market-perceived quality. This can be divided into five factors, from the viewpoint of who can improve them. We will also discuss the methodology for achieving customer satisfaction and gaining advantage in performance fundamentals to increase sales.

8.1   The five factors of market quality

Any corporation should supply products of high value to customers. Having high value means that the product is perceived by the market to have quality relative to its price.[*] Every product has its own selling price, as received by the market.[†] Therefore, needless to say, reducing costs in order to lower prices while maintaining superior product performance is a constant concern of the product development section. Only new technology, as yet unseen over the horizon, will change this equation.

---

[*] See Chapter 2, Section 2.1.
[†] See the beginning of Chapter 5.

Table 8.1 Eight Dimensions and Five Factors
of Market-Perceived Quality

Eight dimensions	Five factors
D1. Performance	F1. Performance
D2. r-Reliability D3. Durability D4. Serviceability	F2. Reliability
D5. Conformance to specification	F3. Conformance quality
D6. Product features D7. Aesthetics D8. Perceived quality	F4. Customer perception
	F5. Fundamentals advantage

If a corporation meets the market price (one of the two elements of product value), its product must have high market-perceived quality. What constitutes quality, as perceived by customers? First, all products should achieve basic quality. Basic quality means that the product provides similar performance as competing products (F1 in Table 8.1), experiences no failure during use (F2), and has no quality defects (F3). No failure means good reliability, and no quality defect means ample conformance to specifications. Maintaining basic quality prevents the market share from plummeting, but does not hike it up. For the product to become a hit, prevailing in the marketplace, it also needs good aesthetics and ease of use (F4), as well as comparative advantages in performance fundamentals (F5). These five factors are shown in the right-hand column of Table 8.1.

Harvard professor David Garvin has examined eight dimensions of product quality.* Although it is difficult to define quality because its meanings are so diverse, Garvin made it easy to understand. He not only differentiated between the failure rate and the reliability lifetime but also clarified quality defects using the concept of conformance to specifications. His eight dimensions include aesthetics and perceived quality, or those things that invite inferences about quality, such as company image or brand, customer use experiences, or assessment reports.

The eight dimensions are listed in the left column of Table 8.1, and a new approach from the viewpoint of how to solve problems in order to improve each dimension is contrasted in the right column. These are the five factors of quality, divided generally according to the types of

---

* D. Garvin, *Managing Quality*, New York: Free Press, 1988, p. 49.

specialists or methods that ensure them. Take television sets as an example. A television reproduces pictures and sound by receiving airwaves (and now, cable communications). Its performance (D1 in Table 8.1) is designed by engineers of electronics and communications. Performance can be considered a first independent factor. Since reliability (D2), according to Garvin, means the failure rate (or rate reliability), and durability (D3) matches the items' lifetime, these two correspond to reliability.* Reliability can be considered a second independent factor because it can be improved by failure identification/analysis technology. Because serviceability (D4) means ease of servicing, and service quality and upgraded reliability minimize service activity, serviceability pertains to reliability. Conformance to specifications (D5) is the task of manufacturing areas. Since it can be improved by manufacturing engineers, conformance to specifications can be considered a third independent factor. Features (D6), aesthetics (D7), and perceived quality (D8) are analyzed by the market research specialist, so these three dimensions can be considered a fourth independent factor.

This reduces Garvin's eight dimensions to four factors, as shown in the right column of Table 8.1: performance, reliability, conformance to specifications, and customer perception. However, there is another dimension of market-perceived quality that cannot be solved with these technologies and methods—comparative performance advantages as sensed by consumers. In order to produce high-quality picture and sound in a TV, the efforts of the electronics engineer are not sufficient. Each performance fundamental of a television—white balance, gray scale, sharpness, and so on—should be positioned to give a comparative edge over competitors. To do that, physicists should understand the product's performance and be able to extract its fundamentals and study their characteristics; electronics engineers should implement their results. In order to secure a differential performance advantage, proceeding one step further than customer satisfaction, specialists from other fields should participate. This dimension differs from the first dimension (performance), so an advantage in fundamentals can be considered a fifth independent factor.

The objectives and related fields of the various factors are arranged in Table 8.2. The top three are necessary conditions for basic quality because the market share of the product can plunge if any mistakes occur at these stages. After the first three necessary conditions are fulfilled, satisfying the lower two increases market share according to the degree of customer satisfaction and advantages in performance fundamentals, so these factors become required conditions.

---

* See Chapter 3, Section 3.1.

Table 8.2 Objectives and Related Fields of the Five Factors

	Factors	Objective (Example: TV)	Related field
Necessary factors for basic quality	1. Performance	Principle of reproducing picture and sound Theory of transmission	Electronic engineering Transmission engineering
	2. Reliability	Analysis and elimination: • Fracture of cabinet • Solder joint fatigue of PCA • Degradation of picture tube	Fracture engineering Mechanics of materials Chemistry Statistics (exponential distribution, Weibull distribution)
	3. Conformance to specifications (c-quality)	Quality inspection system Building the manufacturing process	Statistics (normal distribution) Manufacturing engineering
Required factors preferable to customers	4. Customer perception	Attributes toward product performance and marketing activity	Research in marketing Marketing Aesthetics
	5. Fundamental advantages	Superiority of picture and sound	Physics Basic science

The relationships among the five factors of market-perceived quality are depicted in Figure 8.1. The two factors of reliability and conformance to specifications (excepting performance) are among the factors necessary for basic quality and define quality in the narrower sense.* Reliability pertains to design technology, while conformance to specifications pertains to manufacturing technology. Four factors among the five (excepting conformance to specifications) pertain to design technology. CEOs should understand that making the product successful in the market requires specialists from various fields to cooperate from the first development stage. In order to secure an advantage in fundamentals, it is not enough to rely on existing technology. Since publicized knowledge can be applied to competitors' products as well, new technology and information published openly should be developed and incorporated. Developing unrivaled proprietary technology provides a major market advantage.

---

* As discussed in Chapter 3, Section 3.1.

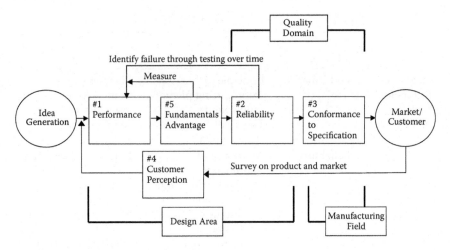

Figure 8.1 Creating world-best product.

8.2   Customer satisfaction

If a corporation views customer satisfaction as a top priority yet does nothing to measure it quantitatively, it is merely shouting slogans and not working seriously on it. And if managers check quality only once and not periodically, they are not managing it. Whatever product issues are targeted, all should be quantified, visualized, and managed continuously. Customer satisfaction begins with measuring satisfaction levels. Customers' opinions on certain questions related to the product can be checked with a questionnaire using a 5-point Likert-type scale. Questions relate to the attributes of the product. The scale indicates the level of satisfaction or importance with the attribute in each question (strongly disagree, disagree, neither agree nor disagree, agree, and strongly agree).

Since product cost increases if all the attributes are satisfied, it is most efficient to satisfy only those attributes considered most important by customers. Knowing the importance level is critical for deciding which attributes should receive attention and improvement. Since attributes should be satisfied according to how important customers believe them to be, shrinking the gap between the surveyed importance and satisfaction should be given the highest priority when considering improvements. It is meaningless to know only the satisfaction level without knowing how important it is to customers. Thus, every question in the survey should be designed to elicit both the importance level and the satisfaction level. With these results, the gap can be calculated. In order to visualize this clearly, think of the importance level and the satisfaction level of a certain attribute denoted as X and Y coordinates, with all coordinates displayed

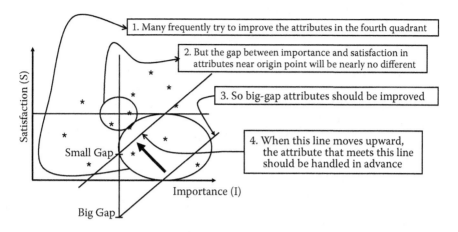

Figure 8.2 Customer satisfaction radar.

in a two-dimensional diagram. New values for the X and Y coordinates can be plotted against the junction of these coordinates, or the average values of the X coordinates and Y coordinates of all attributes (see Figure 8.2).

Many companies try to improve the attributes in the fourth quadrant, because these can roughly be regarded as important but unsatisfactory attributes. But the gap between importance and satisfaction near the junction point is very small, no matter which quadrant the new coordinate is in. Therefore, not all attributes falling in the fourth quadrant are worth an investment in improvement.

The key point is to distinguish all the importance-satisfaction gaps distinctly in the two-dimensional diagram. To do this, the gap is denoted as G and the gap line is defined as $G = X - Y$, or $Y = X - G$, which will be the rightward increase in a 45° line.[*] As depicted in Figure 8.2, the Y intercept of the gap line becomes the gap (G). In other words, the gap for the attributes on the gap line is just the value of the Y intercept. Therefore, the gaps for attributes located on the right-hand side of a certain gap line are greater than those on the left-hand side of the line, and the attributes far below the gap line on the right-hand side are the tasks that should receive top priority.[†] This importance-satisfaction gap analysis is shown in Figure 8.2.

When researching customers' satisfaction with a product, if the satisfaction with competitors' products is also assessed, it is easy to establish a product strategy, because the measurements express the competitive advantages and disadvantages (assuming the competing products have

---

[*] This idea has been initiated by Dr. K. Lee, professor of marketing at Hofstra University (Hempstead, NY).
[†] D. Ryu, *Technology Revolution*, Paju, Korea: Hanseung, 1999, p. 177.

similar prices). According to the model, even though products produce almost equivalent functions, the survey questions and the target customers should be selected using criteria that will vary with the product. Questions about premium products, for example, should be addressed to premium customers. It is impossible to establish a detailed strategy from the data of a large comprehensive survey.

Once a new model is developed, using various tactics to reduce the gaps for important attributes, and is released into the market, the gaps should be measured again to determine whether they have really been reduced or not. Corporations that have close relationships with their customers have their own standardized questionnaires for measuring customer satisfaction. They carry out periodic measurements, watch developments, and analyze improvement activities.

First-class corporations scientifically analyze this kind of data related to customers and keep their questionnaires confidential. Customer satisfaction surveys vary from company to company. Some corporations also survey the initial satisfaction level after product release. They may also assess the satisfaction level of the salesperson as well as the customer, because customer satisfaction is closely connected to the satisfaction of the employees and sales staff who best understand the product. Some corporations also gauge satisfaction with delivery, installation, and service— the so-called moment of truth of the business—to determine issues to be resolved in developing a product strategy.

If a venture company develops a new product, releases it into the market, and successfully secures a bridgehead, customer research programs should still be established in order to sustain competitiveness. It may be wise to consult with a market research specialist regarding customer surveys, survey methods, and survey periods. Some large companies neglect conducting satisfaction surveys and determining attributes for priority improvement, even though people in the company are carrying out satisfaction surveys on service activities. Surveys of customer satisfaction with the product itself should be a top priority, as they express the overall perception of the product by target customers and show the direction in which to go to increase sales. Establishing a customer research system is like installing a radar that tracks customer data at the top of the management chain; the main monitoring screen is the diagram of the importance-satisfaction gap analysis. The beginning of boosting customer satisfaction is surveying the importance and satisfaction levels of the product.

Nowadays, as advances in mathematics and software design have created sophisticated statistical software packages for various fields, market research methods are easy to apply. These methods are very useful tools. If there are more than 100 or 200 respondents to a survey, precious data that could not have been anticipated by anyone in the company are produced.[*]

But this is only true when the researcher clearly sets the target, understands various analysis methods, and properly composes a conceptual framework adequate to the subject being surveyed. The results, using software applied to product design, are automatically generated if the appropriate data are put in as design variables. This requires understanding the constituent elements of the software and the key technology related to product issues. If the software is not providing useful results, it is generally because the necessary data have not been input or some input is insufficient, though sometimes this can be adjusted a little afterwards.

In market research, the composition of questionnaires is the most important stage; it cannot be changed once it is released to respondents. The completed questionnaire will be subject to a statistical software package that processes the data acquired from customers. Thus, the composition of the questionnaire becomes a highly demanding operation. Market researchers should review the conceptual framework of the questionnaire outlined in the written report and the methods for handling data, confirming whether additional questions for unique attributes of the product are needed, or whether the respondents are answering truthfully. How the data will be analyzed should be considered in advance. Bar graphs or pie charts as a summary of survey results illustrate a low level of survey planning, which is caused by insufficiently establishing and reviewing the conceptual framework beforehand. The CEO should review the conceptual framework and analytical methods of the questionnaire before releasing it.

---

[*] Generally, sample size can be calculated with three variables, though this differs depending on the kind of research (G. Ahn, *The Principles of Market Research*, Paju, Korea: Bobmunsa, 1995, p. 226).

$$n = \frac{Z^2 \cdot \sigma^2}{E^2},$$

where E is allowable error (%), σ is standard deviation, and Z is the sample statistic of normal distribution to confidence level (when the confidence level is 95, 90, and 60%, Z is 1.96, 1.65, and 0.84, respectively). The higher the confidence level, the smaller the allowable error, and the greater the standard deviation—then the more the sample size must increase. If it is assumed that the confidence level is 95%, the allowable error of the results stays within 2%, and the standard deviation of the population on a 5-point scale is around 10~13, then the sample size will be about 100. Because market researchers use nonprobabilistic sampling (e.g., convenience sampling) more than probabilistic sampling (e.g., random sampling), sample size will increase more than that calculated with the above equation even under the same conditions. Generally, the required sample size will become around 150 for each product. In the case that the allowable error of results is 4~5% and the standard deviation of the population is 13, then sample size decreases to 5~8 with a commonsense level of confidence, or 60%. Therefore, if CEOs want to confirm whether survey results are adequate and want to get an outline of the situation without detailed research, it would be sufficient to question around 10 target customers.

It is recommended that CEOs study market research methods related to product development—at least roughly grasp their concepts for application—and urge their staffs to apply them to market research planning. The quantification and visualization of surveyed results enhances the accuracy and speed of decision making. These key methods include the following:

1. Importance-satisfaction gap analysis: Visualization via attribute coordinates of importance and satisfaction and the gap between them.
2. Factor analysis: Finding factors, or concealed issues of high importance, extracted from many product attributes.
3. Conjoint analysis: Determining the optimum combination of constituent functions in a product.
4. Multidimensional scaling: Visualization of the competition based on similarities and dissimilarities in products.
5. Correspondence analysis: Visualization of brand position in an image plane.
6. Van Westendorp's price sensitivity meter: Estimation of the appropriate price acceptable to and expected by customers.
7. ServQual survey: Systematic analysis of service quality.
8. Analytic hierarchy process (AHP): Quantitative decision making on various alternatives.

Let's review these various methods individually. The first method, importance-satisfaction gap analysis, has been explained already. If the product attributes scattered meaninglessly in a product survey are analyzed through factor analysis (number 2), they are contracted into several groups.* If all the relationships between any two attributes can be researched through mathematical processing, attributes with high relationships can be assembled into several groups, called *factors*. In other words, the relationships of attributes within a factor are designated as high value and the relationships of attributes in different factors are designated as low value. The name of the factor is determined by interpreting the functions of the attributes within the factor. The importance and satisfaction level of the various factors can also be determined. The factors show the structure one level above that of attributes and reveal in-depth perception patterns in the minds of customers. Because the kinds of perception (factors) and their rankings are critical elements for securing sustained competitiveness, product developers should keep them in mind when designing products.†

---

* J. Lim, et al., *How to Survey Market Research*, Paju, Korea: Bobmunsa, 1997, pp. 141–167.
† M. Kwak and K. Lee, "Facets, Dimensions, and Gaps of Consumer Satisfaction: An Empirical Analysis of Korean Consumers," *Proceedings of the Multicultural Marketing Conference*, Academy of Marketing Science and Old Dominion University, 1996, pp. 324–333.

Conjoint analysis (number 3) is a method providing the optimum combination of product functions or features wanted by customers.[*] In the concept design stage, if several product profiles are defined using product attributes and their levels, and if the preference data for the product profiles are collected and assessed with multiple regression and other methods, then the partial value of levels within attributes can be calculated, as can the optimum product profile with high-value attributes.

Multidimensional scaling (number 4) is a method displaying in multidimensional planes the similarity distance of factors of interest, such as brand or image, based on customer judgments of similarity.[†] On the theory that people understand and memorize complicated relations as simple ways of evading cognitive strain, the market researcher asks about the similarity of products or brands without the kind of complicated questioning involved in, for example, the survey of customer satisfaction. The analysis of similarity data shows the dimension in which customers recognize the objects and where they locate them in that dimension. The visualized result is called a *positioning map*, and its dimension axes can be defined only through data analysis, not in advance. This will be a basic diagnosis, as perceived by the customer, of the competition configuration in the market. As customers compare and prefer products from different angles than the marketers' or developers' views, checking product similarity can be the starting point of creating an idea for a new product.

Correspondence analysis (number 5) transforms a contingency table that is frequently shown as a three-dimensional graph, or divided with many graphs, into data in a two-dimensional plane for easy understandability.[‡] This method was developed later than some of the others discussed here. In our contemporary scientific age, it is possible to transform qualitative images into metric measurements for visualization. Using this technique yields dimensional reduction of multidimensional scaling, similarly to factor analysis and perceptual mapping.[§] This method is the best tool for assessing and improving the aesthetics of a model. Applying correspondence analysis to the relationship between design images and a model (including a competitor's), the image of the model can be identified. For improving images toward what customers desire, this method can also be applied to the relationship between images and design elements to indicate which design elements should be improved.[¶] The result will narrow the design range on which the designer should focus improvement

[*] J. Lim, et al., *How to Survey Market Research*, Paju, Korea: Bobmunsa, 1997, pp. 265–318.

[†] Ibid., pp. 235–261.

[‡] H. Noh, *Theory and Practices of Correspondence Analysis*, Seoul, Korea: Hanol, 2008, pp. 8, 31.

[§] J. Hair, et al., *Multivariate Data Analysis with Readings*, New York: Macmillan, 1992, p. 11.

[¶] K. Shin and S. Park, "Study on the Development for Quantitative Evaluation Method of Product Design," *Journal of Consumer Studies*, 12(4), 2001, 103–118, and D. Ryu, *Technology Revolution*, Paju, Korea: Hanseung, 1999, p. 186.

efforts, which will help ensure that the improved design will be produced speedily and accurately. This method also can be applied to confirming a brand name or nickname emphasizing one feature in a brand. Negative images hidden in a brand name can also be assessed. Sometimes, a proposed nickname thought to convey a positive image can be shown to instead have an unexpected negative connotation, and then be rejected.

One of the major topics in marketing is the confirmation of product price. Obviously, deciding the product price after surveying customers' evaluations of appropriate price is beneficial to sales of a new product or a product with new features, even though a higher price may create a greater profit. If the customers' propensity to pay is surveyed with a price sensitivity meter (number 6) after the new functions and special features are presented to target customers, the difficulties of deciding price can be solved.[*] The questions used in a price sensitivity meter are four comparatively simple items: At what price would you feel that the product is so expensive that regardless of its quality it is not worth buying? At what price would you begin to feel the product is expensive but still worth buying because of its quality? At what price would you still feel the product was inexpensive yet have no doubts as to its quality? At what price would you consider the product to be so inexpensive that you would have doubts about its quality? The responses to these questions can be transformed into four kinds of cumulative distributions of Y axes to the price on X axes, which can reveal the optimal price point and the purchasing intentions of customers as prices change. The information acquired would more closely reflect the market situation than would questions pinpointing a good price or the intention to purchase at several prices checked by customers.

A ServQual survey[†] of service activity has a standard form and has been applied by advanced corporations. This method classifies customer recognition of service activities into five stages (three stages within the organization, from CEO to service personnel, and two stages between the organization and its customers) and clarifies the gap at each stage. In ServQual, these factors of service attributes are reliability (not of hardware, but of service responsibility), responsiveness, assurance, empathy, and tangibles (appearance of physical facilities, equipment, and so on). These factors have been distilled from research into the service provided by industries such as banking and communications in the United States, utilizing factor analysis. Because, with this survey, a program of activities to narrow the gaps can easily be developed and implemented, comprehensive improvement of service quality can be achieved using this method.

[*] K. Monroe, *Pricing: Making Profitable Decisions*, NY: McGraw-Hill, 1990, pp. 114–115..
[†] P. Kotler, *Marketing Management*, 9th ed., Seoul, Korea: Sukjungbooks, 1997, p. 651.

Recently, the analytic hierarchy process (number 8) has been widely applied to various fields and to new product planning.[*] The core concept lies in comparison of pairs in order to judge the importance of all elements—that is, comparing every two elements within the same hierarchical level to each other. AHP was developed on the principle that there is little error in human intuitive power when two things are being compared. The first step in this method is to deconstruct decision-making problems into, let's say, a three-tiered hierarchy—large criteria concepts, smaller concepts, and alternatives that have many elements. Then the relative importance of criteria within the same hierarchical level can be decided by pair-wise comparison. Because different people may assess the relative importance of various criteria differently, and because all criteria are compared in pairs and arranged in order of their importance, the consistency of this hierarchy should be checked with a consistency ratio or consistency index calculated from a pair-wise comparison matrix.[†] Then the relative importance of alternatives can be adduced through pair-wise comparison. Finally, the overall importance of the alternatives is calculated using the importance levels of both the criteria and the alternatives (relative assessment).[‡] If the number of alternatives is great, generally over 10, it is difficult to carry out pair-wise comparison because the number of comparisons increases geometrically. In this case, each criterion should be ranked on a scale of intensity, and the criterion as well as its intensity should be rated in terms of importance using pair-wise comparisons. After this step, each alternative can be given an intensity rank and the overall importance of the alternatives can be calculated using the importance level of the criteria and the importance level of the alternatives' intensity (absolute assessment).[§] This method provides quantitative confirmation of the importance level of the alternatives as well as of the criteria.

There may be other scientific methodologies than the ones discussed here, so CEOs should always be looking for new methods and seeking to understand how their core concepts might be applied to product development. There is also another angle to be considered. In many cases, practitioners in an organization try to find out their bosses' opinions on issues related to customer satisfaction and act accordingly. Actually, the CEO's opinion is just one personal view among many. For example, let's imagine a meeting called to confirm new product aesthetics or a new cookie for children. In this setting, the de facto decision maker is not the president of the meeting, not the salesman, and not even the CEO, but the customers who will purchase the product.

[*] G. Cho, et al., *Analytic Hierarchy Process*, Chungu, Korea: Dhpub, 2005, p. vi.
[†] Ibid., p. 77.
[‡] Ibid., p. 18.
[§] Ibid., p. 39.

These kinds of decisions should be based on market research results from target customers. Thus, it is sensible for CEOs to show interest not in alternatives or conclusions, but in the adequacy of the research methods or processes that provide information about the market situation, and customer responses to the best alternative in order to improve conformance to the market and customer situations. The use of scientific data stops arguments in decision-making meetings, given the many uncertainties of the current business environment. When confirming a new product concept as well as a marketing policy, the basic data come from market research. A product developed without scientific consideration of customer benefits and preferences will naturally be paid little attention by customers, and the root cause of that lack of receptiveness will lie in indifference to scientific methodologies, for which CEOs must be responsible.

8.3   Comparative advantage of performance fundamentals

Where would a product with perfect basic quality hold a competitive advantage? If the plan is to sell it at the same price as similar items, it should easily secure a competitive edge based on core performance, rather than aesthetics or ease of use. No matter how attractive its outside design is, or how easy it is to use, it would be difficult to be competitive if your product is, for example, a compact car that has poor fuel efficiency or a sports car that accelerates slowly. Corporations should gain comparative advantages with their products in the core performance fundamentals.

This can be done by following these steps:

1. Identify the basic elements or the performance fundamentals.
2. Scrutinize and analyze the characteristics of the fundamentals applied to the pilot product.
3. Develop more accurate ways, including obtaining new equipment, to measure these characteristics.
4. Quantitatively measure and evaluate these characteristics and compare them to the results of competitors.
5. Set targets for performance fundamentals and take corrective actions to achieve them.

Comparative advantages are developed using technology appropriate to basic performance issues. With photographic quality, for instance, white point balance and gray scale are critical fundamentals. White balance is the adjustment of the intensities of colors, typically red, green, and blue primary colors, in order to show white in chromaticity coordinates. Since the white balance changes the overall mixture of colors, every television set should, in the assembly process, have the white point set so that color appears natural to the human eye. However, the white point differs from person to person, so it needs to be determined carefully. The chromaticity coordinate of the white point that customers prefer should be measured in a dark room. In order to improve the quality of what is perceived by the five senses, including sight, customer participation is just as crucial as applying the related technology. One interesting result is that the white point changes with eye color. The location of the white point best for people with blue eyes is not the same as for brown-eyed customers. Therefore, it is necessary to change the adjustment standard for products made in Asia from those sold in northern Europe.

Gray scale in the adjustment of a visual display indicates the intensity of luminance adjusted by the magnitude of electric current. The constant increment of current supplied proportionally to the scale mark of the tuning dial produces a sense of sudden brightness at the beginning of dialing and a little change of brightness toward the end of the dial range. This may make customers feel the adjustment dial is not functioning properly. The intensity of the next increment is equivalent, to human senses, to only a fractional increase over each previous intensity. Thus, to make the gray scale comfortable for customers, its adjustment mechanism should be set so that the current increases more as the dial approaches the end of its rotating range. In 1846, Ernst Weber, a German scientist, formulated this theory as an equation, the so-called Weber fraction, which allowed for setting the intensity of luminance in exponential increases to provide the next unit of light apprehended by the viewer.[*] This theory can be applied to the intensity of other human sensations—not only luminance, but the intensity of sound and other sensory perceptions as well.

Some products have well-known performance fundamentals that can be easily deconstructed and analyzed. For example, the performance fundamentals in computers are all determined and their characteristics are well quantified. We can figure out the quality of a computer by checking the performance of the central processing unit, the memory capacity, the data transfer rate, and so on. But because such performance fundamentals vary from product to product, the product developer must isolate and analyze them correctly. Therefore basic scientists, especially physicists and specialists in related fields, must work together to secure competitiveness. CEOs should attend to the importance of technology and the methods intrinsic to their products. While a wide range of technology is required

---

[*] H. Choi, et al., "The Study on Optimization of Gray Scaling in Color Television Picture," *Proceedings of 16th Optics and Quantum Electronics Conference,* Optical Society of Korea, July 1999, p. 212.

for product differentiation or cost reduction, the range of technology scrutinized by scholars is relatively narrow. Therefore, the research chief of the corporation must be able to understand and integrate various fields of technology. It is one of the CEO's tasks to review and absorb new topics explored in various conferences and symposia and to apply them to his products.

Let us take a look at the story of Toyota in the U.S. market. In the late 1950s, cars exported to the United States often failed during highway driving. This was because reliability technology had not been incorporated into their production as a basic quality. At that time there were no highways in Japan, so Toyota did not test long high-speed driving.* After it fixed these reliability issues, Toyota's exports to the United States increased in the late 1960s. Toyota subsequently enlarged its market share in the United States by pursuing competitive advantages in high fuel efficiency and high reliability, while being sensitive to the limits of customer satisfaction.† Recently they intensively studied their premium cars and had been prevailing in the luxury car market, due to accurately evaluating customers' desires and providing them with the technology intrinsic to key issues. At the time of its first release, the price of Toyota's premium car was lower than its competitors', its aerodynamic drag coefficient was lower (0.29), its weight was lower (1,705 kg), its noise and vibration were lower, and fuel efficiency was higher. Toyota relentlessly pursued advantages in almost all performance fundamentals.‡ Unfortunately, in August 2009, it had to pay $10 million to settle a lawsuit over deaths caused by reliability issues that also led to the recall of millions of the automakers' vehicles.§ If there had been no reliability issues and ensuing recalls, Toyota would have continued to prevail in the global auto market.

If a developed product is trouble-free and achieves advantages in major performance fundamentals, how much will its market share increase? That is the subject of Chapter 9.

---

* H. Honda, *The Reliability History*, Tokyo, Japan: Reliability Engineering Association of Japan, 1993, p. 375.

† P. Kotler, *Marketing Management*, 9th ed., Seoul, Korea: Sukjungbooks, 1997, p. 316.

‡ C. Dowson, *Lexus, The Relentless Pursuit* (Korean edition), Seoul, Korea: Keorum, 2004, pp. 107, 137.

§ "Toyota to Pay $10 Million in Crash Suit," *New York Times*, December 24, 2010, p. B7.

chapter nine

Market share increase and market situation analysis

Let's posit that a new product has been developed that has differentiation advantages in its core performance fundamentals and fully satisfies the customer-preferred attributes identified via an importance satisfaction gap analysis. All the reliability issues have been found, corrective actions have been taken according to the results of failure analysis, and the effectiveness of those actions has been confirmed through retesting. Now the market share will remain high and possibly increase. However, it will not increase endlessly. It will finally reach a limit.

Product value increases if a new product improves in quality and is offered at a lower price. The higher the product value, the more the market share increases. If, at this time, the product's market share is estimated, its sales revenue and the ensuing profit could be calculated and the company's CEO could justify the investment adequacy of the new product quantitatively. If the link between product value and market share can be identified, many issues of business management can be resolved. Until now, there has been no theoretical work on this important data. I propose here a new method for estimating the size of the market share increase in advance of releasing a new product into the market.

This estimate can be calculated using the market share of the existing product and the incremental value of the new product over the current one. Applying this equation, we can also estimate the decrease in the number of competitors in the market. With this equation CEOs can understand the extent of the competition and make informed business decisions.

9.1 Predicting the market share increase for new products

The mathematical equation linking product value and market share is both new and simple to apply. The process to derive the final equation seems complicated but can be understood on a commonsense level and is easy to remember. It can also be applied to a newly developed product and to a new product improved by one attribute.

In order to derive the equation for the market share increase due to incremental product value, consider the following example. Assume that the market shares of the products of company A and its competitors are stable and have achieved equilibrium. Company A is about to release a new product, while the competitors' products remain unchanged. The market share of company A will subsequently increase, the competitors' market share will decrease, and the market share of all the companies will eventually reach a new equilibrium. This equilibrium will continue until the release of new products by other competitors.

We know that higher product value (V) earns a higher market share.[*] Because the quantified product value is a comparative figure, customer demand will rush to the higher value product, even if its differences from competitive products are only minor. Therefore, the market share of each company is proportional to the powered product value, defined as a certain power of product value (V^α). The constant α becomes a value multiplier in the market and can appear differently according to the product market or the quantification method. The larger the value multiplier, the more a small increment in product value will be magnified to a greater market share increase. Because the value multiplier will vary according to the products and their market situation, understanding comparative features as perceived by customers will increase the value multiplier.

The relation between the market share (S) and the powered product value (V^α) can be expressed as follows:

$$S \equiv K \cdot V^\alpha \tag{9.1}$$

where K is a constant proportional to the market share, V is product value, and α is a value multiplier greater than 1.

Product value (V) can be expressed as a function of two variables, price and quality:[†]

$$V = f(P, Q) \tag{9.2}$$

[*] B. Gale, *Managing Customer Value*, New York: The Free Press, 1994, p. 82.
[†] As discussed in Chapter 2, Section 2.1.

If all products are at the same price level, the product value will be the perceived quality profile, as follows:[*]

$$V = Q = a_1 \cdot Q_1 + a_2 \cdot Q_2 + \cdots \tag{9.3}$$

where Q_1, Q_2 are the satisfaction scores of the quality attributes, and a_1, a_2 are the importance scores, or weight, of the quality attributes.

As the sum of the market share of all companies becomes 1, the proportional constant of Equation (9.1), K, can be figured out. Then the definition of market share will be as follows:

$$S = K \cdot V^\alpha = \frac{1}{V^\alpha + V_1^\alpha + V_2^\alpha + \cdots} \cdot V^\alpha \tag{9.4}$$

where S is the market share of company A, V^α is the powered product value of company A, and V_1^α, V_2^α are the powered product values of competitors 1 and 2, respectively. Ultimately, the market share of company A is the ratio of the powered product value of company A to the sum of the powered product values of its competitors, as well as company A.

[*] Considering the product price, product value is quantified as a linear equation (B. Gale, *Managing Customer Value*, New York: The Free Press, 1994, p. 128):

$$V = a_1 \cdot Q_1 + a_2 \cdot Q_2 + \cdots + b_1 \left(\frac{1}{P_1} \right) + b_2 \left(\frac{1}{P_2} \right) + \cdots$$

where Q_1, Q_2 are quality attributes, P_1, P_2 are price attributes, and a_1, a_2, b_1, b_2 are their coefficients. But expressing this with a ratio equation is easier to apply practically than expressing it with a linear equation. Thus, I propose that product value (V) should be defined as the ratio of the perceived quality profile over the specific price (p), as follows:

$$V = \frac{Q}{p^\gamma}$$

where p is the specific price and γ is the price multiplier. If γ is greater than 1, the product value will greatly change due to its price. The specific price (p) is the ratio of the perceived price profile to the average perceived price profiles (P_{avr}) in the market as follows:

$$p = \frac{P}{P_{avr}}$$

The perceived price profile (P) is

$$P = b_1 \cdot P_1 + b_2 \cdot P_2 + \cdots$$

where P_1, P_2 are price attributes scores, and b_1, b_2 are the coefficients of attributes. The average perceived price profile (P_{avr}) is the average of the perceived price profiles of all competitors. For price attributes, refer to B. Gale, *Managing Customer Value*, New York: The Free Press, 1994, p. 45.

In order to quantify the value increment of company A's new product over the existing product, let's introduce a new variable, called the *powered value increment (R)*:*

$$R \equiv \frac{V_N{}^\alpha}{V_P{}^\alpha} - 1 \qquad (9.5)$$

where $V_N{}^\alpha$, $V_P{}^\alpha$ are the powered product values, respectively, of company A's new and existing products.

The market share of new product (S_N) is then

$$S_N = K_N \cdot V_N{}^\alpha = \frac{1}{V_N{}^\alpha + \left(V_1{}^\alpha + V_2{}^\alpha + \cdots\right)} \cdot V_N{}^\alpha \qquad (9.6)$$

where K_N is the proportional constant of the new product to the market share.

Now let's derive the market share increase (ΔS) due to the new product, using Equation (9.6). First, substitute known variables for the two unknown variables in Equation (9.6)—the powered product value of the new product of company A, $(V_N{}^\alpha)$, and the sum of the powered product values of its competitors, $(V_1{}^\alpha + V_2{}^\alpha + \cdots)$. The market share of the existing product (S_P) is as follows:

$$S_P = K_P \cdot V_P{}^\alpha = \frac{1}{V_P{}^\alpha + \left(V_1{}^\alpha + V_2{}^\alpha + \cdots\right)} \cdot V_P{}^\alpha \qquad (9.7)$$

where K_P is the proportional constant of the existing product to the market share. Second, the sum of the powered product values of the competitors, $(V_1{}^\alpha + V_2{}^\alpha + \cdots)$ is derived from Equation (9.7) as follows:

$$\left(V_1{}^\alpha + V_2{}^\alpha + \cdots\right) = \frac{1}{K_P} - V_P{}^\alpha \qquad (9.8)$$

---

* In order to extract a variable appropriate to the market share increase (ΔS), we can apply the following equations:

$$\Delta S = S_N - S_P = K_N \cdot V_N{}^\alpha - K_P \cdot V_P{}^\alpha = K_\Delta \cdot V_P{}^\alpha \cdot \left(\frac{V_N{}^\alpha}{V_P{}^\alpha} - 1\right)$$

where S_N, S_P are the market shares of the new product and the existing product, K_N, K_P are their proportional constants, $V_N{}^\alpha$, $V_P{}^\alpha$ are the powered product values of the new and the existing products, and K_Δ is a constant proportional to the market share increase. The last term is extracted from the variables of the powered value increment (R). For reference, the relations of constant, K_N, K_P, K_Δ, are as follows:

$$K_N = K_P \cdot \frac{1}{1 + S_P \cdot R}, \quad K_\Delta = K_N (1 - S_P) = K_P \cdot \frac{1}{1 + S_P \cdot R} \cdot (1 - S_P)$$

Finally, the powered product value of company A's new product, (V_N^α), is derived from Equation (9.5):

$$V_N^\alpha = (1+R)\cdot V_P^\alpha \tag{9.9}$$

Substituting these two equations, 9.8 and 9.9, into Equation (9.6), the market share of the new product (S_N) is derived as follows:

$$S_N = S_P \cdot (1+R)\cdot \frac{1}{1+S_P \cdot R} \tag{9.10}$$

Therefore, the market share increase (ΔS) due to the introduction of the new product is[*]

$$\Delta S = S_N - S_P = (1-S_P)\cdot S_p \cdot R \cdot \frac{1}{(1+S_P \cdot R)} \tag{9.11}$$

With the two known data, or the market share of the existing product (S_P) and the powered value increment (R) of the new product over the existing product, we can calculate the market share increase (ΔS) due to the introduction of the new product. The market share increase (ΔS) will take up the portion of the remaining attack target $(1 - S_P)$ multiplied by the current market share (S_P) and the powered value increment (R). Since the sum of the market share of all the companies is 1, the last term of Equation (9.11) will be the coefficient for adjusting the sum to 1. The reason for solving the equation with two variables is that the current market share corresponds to the competitive edge of the existing product, marketing competence, service activities, and so on.

Fitting the equation for the powered value increment (R) to the following series of equations confirms the meaning of the value multiplier (α) and its adequacy:

$$R \equiv \left(\frac{V_N}{V_P}\right)^\alpha - 1 = \left(1+\frac{\Delta V}{V_P}\right)^\alpha - 1 = \alpha \cdot \frac{\Delta V}{V_P} + \frac{\alpha \cdot (\alpha - 1)}{2}\cdot \left(\frac{\Delta V}{V_P}\right)^2 + \cdots \tag{9.12}$$

where V_N equals $V_P + \Delta V$, V_N, V_P are the product values of the new and existing products, respectively, and ΔV is the product value increase. When the ratio of the product value increase to the existing value $(\Delta V/V_P)$

[*] D. Ryu, "Quality, Product Quality, and Market Share Increase," *International Journal of Reliability and Applications* (English), 2(3), 2001, 179–186.

is sufficiently small, the powered value increment (R) and the market share increase (ΔS) will be approximated to linear relationships as follows:

$$R \cong \alpha \cdot \frac{\Delta V}{V_P} \tag{9.13}$$

$$\Delta S \cong (1 - S_P) \cdot S_P \cdot \alpha \cdot \frac{\Delta V}{V_P} \tag{9.14}$$

The market share increase (ΔS) is proportional to the current market share (S_P), the value multiplier (α), and the ratio of the product value increase over the existing product's value ($\Delta V / V_P$), and is confined to the sum of the competitors' market shares ($1 - S_P$). Tiny improvements in the current product will make the market share increase greatly due to the value multiplier.

Now let's estimate the market share increase for an example case. For a certain retailer, the current market share of company A (S_P) is 17.6%, and those of two competitors (S_1, S_2) are 35.3 and 23.5%, respectively; the remainder is accounted for by small companies. In order to overcome its inferior market position, company A develops a new product and surveys the level of customer satisfaction about both its and its competitors' products. Since all the products are at the same price level, the perceived quality profile will be the product value. The satisfaction scores of all the attributes are summed with a weighted rating (the importance scores of the quality attributes). The product value of company A (V_N) is calculated as 65.7, and the product values of competitors (V_1, V_2) as 64.8, 62.2. These values are obtained by converting the importance and satisfaction scores on the 5-point scale in the customer satisfaction survey into percentiles as follows:[*]

$$V = a_1 \cdot Q_1 + a_2 \cdot Q_2 + \cdots = 100 \cdot \frac{1}{n} \cdot \frac{1}{5 \times 5} \cdot \left(i_1 \cdot s_1 + i_2 \cdot s_2 + \cdots \right) \tag{9.15}$$

where n is the number of attributes, i_1, i_2 are the importance scores, and s_1, s_2 are the respective satisfaction scores. As the product value of company A is better than that of its competitors, the market share of company A will be estimated to increase. The value multiplier (α) can be obtained using Equation (9.1):

$$\frac{V_1^\alpha}{V_2^\alpha} = \frac{S_1}{S_2}, \text{ then } \left(\frac{64.8}{62.2} \right)^\alpha = \frac{0.353}{0.235}, \therefore \alpha = 9.94$$

[*] As the number of attributes is n and the Likert scale uses 5 points, the sum will be divided by $(5 \times 5) \cdot n$.

And though the product value of the existing product was not surveyed, it can be derived as follows:

$$\frac{V_P^{\alpha}}{V_1^{\alpha}} = \frac{S_P}{S_1}, \quad V_P = V_1 \cdot \left(\frac{S_P}{S_1}\right)^{\frac{1}{\alpha}} = 64.8 \times \left(\frac{17.6}{35.3}\right)^{\frac{1}{9.94}}, \quad \therefore V_P = 60.4$$

The powered value increment (R) of the new product over the existing product can be calculated by Equation (9.5) as follows:

$$R = \frac{V_N^{\alpha}}{V_P^{\alpha}} - 1 = \left(\frac{65.7}{60.4}\right)^{9.94} - 1 = 1.299$$

This figure represents not just a simple improvement, but a quantum leap. The market share increase (ΔS) and the market share (S_N) of company A can be estimated to be in a new equilibrium by Equation (9.11):

$$\Delta S = (1 - S_P) \cdot S_P \cdot R \cdot \frac{1}{1 + S_P \cdot R}$$

$$= (1 - 0.176) \times 0.176 \times 1.299 \times \frac{1}{(1 + 0.176 \times 1.299)} = 0.153$$

$$S_N = 0.176 + 0.153 = 0.329$$

The market share of company A will increase by 15.3%—to 32.9% from 17.6 %—and that of the two competitors will decrease to 28.7 and 19.1%, with a shrinkage of 6.6 and 4.4%.* This case actually occurred in my experience, and the market share increased from 17.6% in January 2000 to 24% in July 2000 due to introduction of the new product. The reliability and basic quality of the new product had been perfected and a competitive edge had been achieved in performance fundamentals.† When I posed as an anonymous shopper eager to buy a competitor's product in one retail outlet, the salesman enthusiastically tried to persuade me of the superiority of company A's product. Good products make convincing salesmen.

When one attribute is improved, we can estimate how much the market share will increase using the same process. For example, assume that the importance and satisfaction scores (I, S) of service quality, one of the attributes in the customer satisfaction survey, increased from 4.5, 3.5 to

* $S_{1N} = 32.9 \times \left(\frac{64.8}{65.7}\right)^{9.94} = 28.7$, $S_{2N} = 32.9 \times \left(\frac{62.2}{65.7}\right)^{9.94} = 19.1$.

† See Chapter 8, Section 8.3.

4.5, 4.0, respectively. In this case, let's estimate the market share increase (ΔS) and the market share (S_N). The current market share (S_P) is 17.6%, the percentile of the product value (V) as surveyed is 60.4, the number of attributes of customer satisfaction surveyed is 20, and the value multiplier (α) is 9.94.

The product value increase (ΔV) is calculated using Equation (9.15) as follows:

$$\Delta V = 100 \cdot \frac{1}{n} \cdot \frac{1}{5 \times 5} \cdot (I' \cdot S' - I \cdot S)$$

$$= 100 \times \frac{1}{20} \times \frac{1}{5 \times 5} \times (4.5 \times 4.0 - 4.5 \times 3.5) = 0.450$$

(9.16)

Thus, the product value (V_P) increases from 60.4 to 60.85, and the ratio of the product value increase to the existing product value ($\Delta V/V_P$) is 0.45/60.4, or 0.745%. As the ratio of the product value increase is small, the market share increase can be estimated with Equation (9.14):

$$\Delta S \cong (1 - S_P) \cdot S_P \cdot \alpha \cdot \left(\frac{\Delta V}{V_P} \right) = (1 - 0.176) \times 0.176 \times 9.94 \times \frac{0.45}{60.4} = 0.01074$$

If product reliability improves, the product value increases as the satisfaction scores grow and repair expenses will be reduced directly. Consequently, sales revenue increases and market share rises by about 1.1%. Note that the 1% increase in market share is not small, because the 0.5 increase in the satisfaction score is for an attribute with an importance score of 4.5.

Using a customer satisfaction survey to research product value is extremely useful for estimating the market effect of improving various attributes and subsequently prioritizing them. To gain a competitive advantage, various market research methods and related technology intrinsic to the attributes should be mobilized.

9.2 Extreme competition until two strong corporations emerge

We have established that the market share changes in accordance with the powered product value (V^α). Therefore, when a company enlarges its market share with new products that have improved functions, competitors should develop similar products and chase after the leading company. If a company neglects activity like this, it will fade out of the market. It is difficult to lead the market with new, fresh products and also difficult to chase the leading company.

Let's consider a competitive situation. Assume that the market shares of several companies are all the same because they are selling quite similar products of equal product value. Assume also that all the companies but one are striving to improve product value similarly with respect to one another—that is, they have an equal powered value increment (R). How great a powered value increment will make the company neglecting development (company X) fade out?

The decline of company X occurs when the sum of the market share of the improving companies exceeds the market share of company X, as in the following equation:

$$S_0 \le (n-1) \cdot \Delta S \tag{9.17}$$

where S_0 is the market share of company X, n is the number of competitors, and ΔS is the estimated market share increase due to the introduction of new products. Substituting Equation (9.11) for the market share increase, ΔS of Equation (9.17), we obtain

$$S_0 \le (n-1) \cdot (1-S) \cdot S \cdot R / (1+S \cdot R) \tag{9.18}$$

where S is the market share of the innovative companies and R is the powered value increment of the new products. Since the market shares of all the competitors are equal, then $S_0 = S = 1/n$. Solving this equation for the powered value increment, we get

$$R \ge \frac{1}{n-2} \tag{9.19}$$

Let's consider the meaning of these results. When five out of six companies endeavor to increase product value, their powered value increment (R), 25%, would be sufficient to push company X out. A powered value increment of 25% in product improvement seems great but actually is not. The value multiplier (α) is not small—it is about 10 in the example in the previous section. If the value multiplier is 10 and the powered value increment is 25%, the ratio of the product value increase to the existing product value ($\Delta V/V_P$) will be about one-tenth, or 2.26%—not very high, as imagined with approximated Equation (9.13).

When four of five companies increase their product value, the ratio of the product value increase ($\Delta V/V$) of 2.89% for the four companies would be enough to defeat company X. When three out of four companies are

Table 9.1 Situation Analysis of Competition

Product value increase of (n–1) companies pushing out company X					Remarks
Number of companies (n)	7	6	5	4	3	
Powered value increment (R) (%)	20	25	33	50	100	
Ratio of product value increase ($\Delta V/V$) (%)	1.84	2.26	2.89	4.14	7.18	Value multiplier; $\alpha = 10$
Product value (V) (percentile scores of customer satisfaction)	61.5	61.8	62.2	62.9	64.7	Baseline of customer satisfaction; 60.4

improving, the ratio of the product value increase—4.14%—would be enough for company X to be pushed out. In the example in the previous section, the ratio of the product value increase (4.14%) corresponds to the change in the percentile scores of customer satisfaction from 60.4 to 62.9—a 2.5-point rise. This activity improving the satisfaction degree seems difficult but is not impossible to attain. Therefore, improved products should be released in advance of competitors' products, and if a competitor releases a new one, it is important to catch up rapidly. Thus if a company wants to remain in the leading group, it must incessantly watch competitors' activities. Since competitors are in a neck-and-neck race, CEOs must devote themselves to their businesses ceaselessly. In the case of seasonal goods, the existence of a fixed release date lightens the CEOs' burdens. All this is summarized in Table 9.1.

Assume that three companies among six are falling behind in new product competition and three companies remain. In order to eliminate one of the slow-moving companies from the market, the other two need to increase the powered value increment (R) by 100%, or 7.18% of the ratio of the product value increase ($\Delta V/V$). Such an improvement is not easy because it requires a quantum leap in product value. This might not happen often, but it can and does occur. In addition, although pushing out one company of two roughly competitive companies would not be impossible, two strong ones could perhaps coexist for another reason.

In the 1910s, there were more than 200 companies manufacturing cars in the United States. The number of carmakers was only 20 in the 1930s, 4 by the 1960s, and as of now, 3.[*] Before the global financial crisis of 2008, the leading two had been profitable and the third appeared to be

---

[*] S. Yoon, *Systemic Theory of Management*, Seoul, Korea: GyeongMoon, 2002, p. 215.

sometimes in deficit, sometimes in surplus. In a global competition without trade barriers, is it unrealistic to estimate that about 10 carmakers in the world would be decreased to 2 or 3 in every product segment?

Imagine that there would be two strong companies for every product worldwide. The competition among them would be fierce, so CEOs should never be careless. Actually in the business of basic components and materials, such situations have already occurred. Extreme competition will continue until two strong companies remain, except possibly in some recently generated high-tech fields.

9.3   Some suggestions for competitive capability building

The discussion so far seems incomplete without considering how to enter the leading group, even though it deviates from our main theme of the product.

Over the past quarter of a century, the major developed countries—what Chang calls the "bad Samaritans"—using the unholy trinity of the International Monetary Fund (IMF), the World Bank, and the World Trade Organization (WTO) (naturally, including General Agreement on Tarrifs and Trade (GATT)), and so on, have made it increasingly difficult for developing countries to pursue appropriate development policies.* This is because these systems do not allow underdeveloped countries enough time to enter the industrial orbit. As mentioned in the previous section, no matter how the market grows, the number of competitors will remain roughly the same or be reduced to a few, regardless of the market size. This means that low-tech corporations have rather less opportunity than when trade barriers existed, and that the best products developed with advanced technology by the leading companies will be the ones to survive in the global market. In order to enter the market, any corporation must develop good products that surpass, or at least equal, the products of top companies. A new product should be perfect from the first to acquire a good portion of the market share, using its own competitiveness without any supplementation—like a newly fledged albatross flying high because of its innate and hereditary ability, without any instruction or the need for a period of trial and error.

As everyone understands, the repeated creation of good products is not only a matter of technology. The fact that good products are favored in the market when they are released is the result of the creative and efficient activities of the corporate organizations generating such products. If organizations hold an advantage in operation, their products never are

---

* H. Chang, *Bad Samaritans*, Seoul, Korea: Bookie, 2007, p. 329.

conquered in the market. Although this involves a wide range of issues, I propose some suggestions for cultivating the kind of creative organizations that dominate the market.

Corporate competition has changed to system competition and, further, to ecosystem competition, as mentioned in the first chapter. Among corporations with the dynamic capability of responding properly to changing circumstances, only those that have been developing organizational capability can survive. Superiority in organizational competence and personnel skill results in good products. All the companies in each field are competing to build capability.

A corporation is not just an aggregate of individuals, and a large enterprise is not simply an assemblage of small ones. Let's deconstruct organizational competence in terms of "corporate DNA" adapting and evolving under the competition in an ecosystem. Living things have two kinds of traits—inherited traits and adaptive traits.[*] Inherited traits are intrinsic characteristics that are not influenced by the external environment; adaptive traits are acquired features better suited for survival in the outer environment. For corporate survival, inherited traits are necessary conditions and adaptive traits are sufficient conditions, though it is not easy to differentiate the two.

It is said that the characteristic of a successful long-lived company is that it has stored a "previously learned pattern of action" in its corporate DNA or in corporate documents that are transmitted to the next generation of personnel.[†] This implies that people in successful long-lived companies are working more systematically than those in less successful companies. What does this mean specifically?

When we visit an advanced corporation, we are immediately impressed, from the main entrance to the well-arranged buyers' meeting rooms to the helpful receptionist. Do these competences and the organizational structure of advanced corporations reflect corporate DNA?

We can identify the corporate DNA by asking a series of questions and getting positive answers. Will nicely furnishing a meeting room for buyers help corporations produce world-best products? Will organizing the structure of a company in the way advanced corporations do make the company produce first-class products? No. There are deviations in corporate DNA, many of them of superficial importance. Let's suggest some questions to determine what the intrinsic corporate DNA is.

---

[*] K. Kim, "Comparative Study on the Evolution Vector of Business Architecture between Korea and Japan," *The Korean Small Business Studies*, 27(3), 2005, p. 160.
[†] Ibid., p. 162.

1. When a negative incident occurs suddenly, are the related personnel assembled quickly? Is the incident handled speedily? Is the plan for responsive activities adequate?
2. At the beginning of every month, are the production data of the last month reported? Is the current situation understood quickly and correctly? Are the key issues identified and reported immediately to the appropriate people?
3. Is the probability of fixing an issue high because solutions are selected according to the results of in-depth and scientific analysis?
4. Is the decision making on critical pending issues determined by the opinions of one or several executives or by scientific and quantitative data?

Now let's consider the stages of evolution of corporate DNA by reviewing three alternative behaviors:

1. When a subject is very critical and the decision maker is clearly the CEO, related personnel wait for the CEO's decision.
2. Preparing for an anticipated meeting about the problem, personnel are already considering possible solutions related to their fields.
3. The relevant person is reporting some solutions and alternatives and their merits/demerits to the CEO, helping the CEO to gather the data necessary for making the decision.

We can infer that the third organizational pattern is the most successful.

I personally think that the core inherited trait in corporate DNA is the modus operandi of the participants in a meeting. Organizations are constituted specifically to surmount the limits of individual ability, and the synergy produced by bringing together relevant personnel greatly influences corporate success. Operating meetings in successful long-lived companies are much different than those in ordinary companies. Let's evaluate the level of inherited traits in corporate DNA by reviewing five kinds of operating meetings:

1. Personnel in the meeting do not know the reason for the meeting, or do not acknowledge the issues being addressed. The purpose of the meeting is to create awareness of the issues.
2. Personnel at the meeting intuit vaguely what the issues are. There are many impromptu remarks at the meeting that evoke organizational conflict afterward.
3. Someone already has finished reviewing and analyzing the issues. This person explains what is going on systematically and makes sure members recognize the issues. Action items and the date of the next meeting are decided before closing the meeting.

4. Some of the people at the meeting recognize the issues and are prepared to present solutions; the others are unprepared and simply respond. Solutions are superficial and it takes a lot of time to draw conclusions.
5. All personnel recognize the issues and propose alternatives for adequate solutions. An activity plan is constructed using their ideas.

In successful long-lived companies, the personnel in a meeting, their state of preparedness, and the operation of the meeting are advanced. In other words, the meeting groups—the marketing section and product development section, or the product development section and manufacturing section, and so on—are adjusted to match the size of the corporation. CEOs pay attention to the process of being induced to draw the best conclusion rather than drawing their own conclusions. Meetings are held periodically and people in the organization move organically—that is, the cells work without error because they have highly evolved DNA.

Successful long-lived companies also have well-evolved work systems, in which rules and work processes are advanced. Systems for reviewing ideas generated both internally and externally are different from those in less evolved companies. So are the methods for sharing knowledge in the community of practice to improve the capability of individual staffers, and the system for researching best practices and disseminating them. The mechanisms for absorbing new technology also differ. For example, top companies absorb appropriate new technology by participating in technological conferences in related fields as well as attending new product exhibitions. In order to evaluate technological advancement in the corporation, an annual review evaluates the hiring of specialists, acquired technical materials, and newly developed equipment.* At the end of every year, the financial statements and business performance are clarified, the business strategy of the next year is publicized, and a 3-year rolling plan is also set. The CEO realizes that the business confronts global competition and inspires people to remain forward thinking and enthusiastic. Furthermore, in order to create a sense of unity among all personnel, the CEO proclaims both "vision and will" so that employees will continually develop their knowledge and ability and realize their ambitions inside the company.

So far, we have mainly discussed the inherited traits of corporate DNA, but what about the adaptive traits? They correspond to the speed of the response to competitors' new products, the ability to reduce costs for current products, the managing capability of ad hoc committees to handle issues, and so on.

Likewise, many aspects of organizational competence should be continuously improved in order to maintain a comparative winning edge

---

* See Chapter 1, Section 1.2.

over competitors. Even more important than organizational competence is competence in idea generation. Idea creation without any trade-off—for example, producing high performance without a cost increase—is a prerequisite for solving almost all problems. The idea creation process consists of five stages.* Initially, an individual discovers that something is worth working on or becomes aware of a problem, which is the first stage—problem finding or sensing. Then, this person concentrates on the problem and collects relevant information without, at this point, any refinement. Accumulating information will lead to the right conclusions. All relevant information should be researched in detail and gathered together, which is the second stage—immersion in preparation. Next, this person internalizes as much of the data as possible and mulls it over in the back of his mind, or even forgets about it for a while, which is the third stage—incubation or gestation. Actually, during this stage the information is not really forgotten; the brain works incessantly, searching for solutions, even at a subconscious level. At an unanticipated time when the brain finishes computing, a new idea flashes into the individual's mind. This is the final result of "backstage" mental computation, the fourth stage—insight or illumination. Finally, the person sets out to prove by logical argument or experimentation that the creative solution has merit during the final stage—verification and application.

If, over considerable time, no new ideas are generated, this indicates incomplete preparation. In other words, although the brain is still working, it cannot create output in the form of creative solutions because insufficient information has been input. Thus, scrutinizing the situation in detail, classifying the acquired information, and learning it thoroughly are critical to idea creation.

Let's take a closer look at this. When organizational competence is evaluated, the comparative advantages of many items are also reviewed. For a given product, for instance, indices such as product performance, development expenses, the time necessary for product development, manufacturing productivity for the item, product quality, and so on, are applicable. Although there are necessary activities related to resolving the difference between the targets for each characteristic and the status quo, all the indices mentioned are results indirectly indicating the level of evolution of the corporate DNA. To evolve even further, CEOs need to examine the basis of all the corporate activities.

Some frequent tasks are somewhat superficial and reflect only the resulting phenomena and, therefore, are not themselves analyses and measures of basic causes. For example, an efficiency meeting should be completed in an hour, or 2 h at most. Advance preparation for the meeting,

---

* A. DuBrin, *Foundation of Organizational Behavior*, New York: Prentice-Hall, 1984, p. 200, and D. Ryu, *Corporation Revolution*, Seoul, Korea: Hanseung, 1997, p. 215.

development of an agenda, the remarks made by staff members, the unequal distribution of comments among personnel, the process of drawing conclusions, the minutes, and so on, reveal many factors that could be improved in advance to keep meetings relatively short and effective.

Various topics related to organizational competence, like the relevance of staffers' activities to their assigned responsibilities, cooperation between younger and more experienced staff, and the fostering of creativity generation, should be addressed. Examples of significant questions might include: Will someone who comes to work on Friday already dressed for a weekend trip be attentive to the tasks at hand that day? Could the old veteran chief in manufacturing be irritated by orders handled poorly by the young salesman, rather than seeing it as an opportunity for training? Will the successful development chief reject out of hand the ideas of an ambitious new person? Recognizing and addressing such issues will lead the CEO to understand that some corporate DNA might be hidden or gradually degrading; the CEO can then concentrate on promoting more positive practices.

Considerations like these should not be applied only to organizational competences. Individuals in the organization should also be considered carefully. For example, the CEO needs to have confidence that the product development chief is more knowledgeable and a better personnel manager than his or her opposite number in competing companies. The superiority of individual capability and organizational competence will always produce world-best products.

Conclusion

If a new product is designed with new materials or a new structure, approval specifications for it should also be addressed anew. This is a universally applicable approach. Just because a set of specifications is well established does not mean they remain valid regardless of changed circumstances. Any change requires revising the specifications, sometimes quantitatively. This also pertains to design functions; otherwise, reliability accidents related to poor quality will occur at some time even in advanced corporations.

Reliability technology involves estimating the future behavior of hardware products after 10 or 20 years. It is difficult but not impossible to do this. Accurately grasping reliability concepts can motivate an organization to adopt scientific principles, minimizing quality problems down the line.

In order to enter the rank of leading corporations, the CEO's paradigm should be totally revised. Many factors at the early stage of setting the initial product concept should be reexamined and revisioned, from whether to outsource specifications or handle them internally to whether test equipment should be purchased or designed and made inside. The entire process should be developed based on a keen understanding of the intrinsic technology. Trying to copy other companies' products without understanding the underlying principles is an inferior strategy that will lead to an inferior product. Performance problems, as well as reliability issues, should be deconstructed into subunits, and for each of these, performance should be investigated and measured quantitatively. Customer needs and desires should be researched not by simple frequency studies but by advanced methods producing diagrammatic visuals, and should be confirmed periodically in depth and multilaterally.

Providing the organizational competencies to incorporate these methods will allow the firm to join the group of leading producers. Changing basic thinking is the required condition, and the synthesized results will lift corporate success.

Multiple competitors in the leading group are in an unstable transition during situations like the early competition over a new product, as we have demonstrated quantitatively. In the long run, top competitors in all fields will be reduced to two or three strong companies in a confined market. The economic war waged by powerful Western companies recently will remove all kinds of local competition and eventually become a cold war between the two strongest corporations. Just as the United States and western Europe struggled against the USSR after World War II and, more recently, U.S. and Japanese corporations fought economic competition in Europe, the strongest entity will share the global market with the next strongest. Even after entering the leading group, a CEO should improve the company ceaselessly until only two strong companies remain, by encouraging and implementing new and unique ideas and strengthening management principles.

It is not easy to prevail over the market competition. The core performance of a product must be good in relation to its price. Closely watching customer preferences will indicate the success of a product's core performance and related features. The intrinsic technology and relevant new ideas will solve issues determined in the market.

The intrinsic technology will advance as an understanding of the basics deepens, which is the task not only of staff but also of the CEO. The CEO's encouragement should be reflected in detailed observations and remarks, not simply in cheery greetings. All too often, CEOs do not understand staff explanations of problems and methods, not because the CEO is insensitive or ignorant, but because the staff do not themselves understand them fully and cannot explain them in commonsense terms. Thus, it is also critical for the CEO to request staff members to have a deep and firm grasp of fundamentals.

The generation of ideas or insight results from understanding, memorizing, and synthesizing common sense. Just as a computer can only generate good results when enough accurate data have been input, a lack of new ideas in a corporation indicates a lack of commonsense thinking or of required data. If commonsense understanding and adequate information accumulate in the brain's computer, new ideas will be generated; the more data are input to the brain, the more abundantly excellent ideas will begin to flow.

If a product operates trouble-free, is easy to use, and demonstrates a comparative performance edge over similarly priced competing items, it will succeed in the market and its producer will survive and flourish.

Appendix: E-mail to CEO regarding reliability management

These excerpts from the book summarize the key concepts that differentiate reliability technology from quality control. They focus on the concepts of reliability technology and its ensuing tasks for CEOs and directors who do have not enough time to read the entire book.

A.1 The operation of a reliability organization is a must in corporate hardware development

Several years ago, a medium-sized company regarded as a rising star in Korea declared bankruptcy. It had developed mobile phones and launched them successfully in the domestic market and then enlarged overseas. Due to price competition in China, its profitability had deteriorated, so in the expectation of higher prices, it opened up a new market in Europe. But there was also severe price competition in the market and the company released its product without enough margin. Unfortunately, due to an inherent quality problem, the returned products piled up, which forced the business to close, finally, with major losses. In the World Trade Organization (WTO) system converted from the General Agreement on Trades and Tariffs (GATT) in order to enhance free trade, the global marketplace has become increasingly competitive; companies can only survive when they make no mistakes in either price or quality.

Table A.1 Business Activities Necessary to Remain Competitive

	Meeting customers' requirements	Maintaining an efficient business
Purpose	Customer satisfaction	Realizing surplus (creating profit)
Details	1. Basic quality (no defects or failures) 2. Good aesthetics and ease of use 3. Performance edge	1. Design for reduced material and manufacturing costs 2. Reduce development period and expenditures 3. Reduce marketing, production, and shipping expenditures 4. Conduct research and development for continuing competitiveness

There are two kinds of goals to be pursued in the business of making hardware products, as shown in Table A.1. One is to meet customers' requirements; the other is to maintain an efficient business. Meeting customers' requirements means supplying good products. First, the product should fulfill basic quality standards, meaning that it performs like competing products, that there are no defects, and that it does not fail during use. But a product with just this basic quality does not attract customers' attention and therefore can only be sold at a cheaper price; under these conditions, it is hard to ensure a profit. In order to receive customers' favor, a product should display good aesthetic design, ease of use, and special comparative features in performance.

Maintaining good business principles means making profits an ongoing concern—the result of four kinds of activities that give a manufacturer an edge over the competition. First, a product should be designed so as to reduce material and manufacturing costs. Second, the product should be developed in a shorter period and less expensively than those of competitors. Third, it should be produced, introduced, promoted, and conveyed to customers efficiently. Finally, various technologies relevant to the product should be studied and researched to ensure a continuing stream of high-quality future products.

Product design, the first of these features, has a great influence on realizing surplus. Designing a product to be cheaper is not easy, but is a constant goal in hardware manufacturing. The product should be designed to be as dense and compact as possible to meet customers' requirements. As the reduced material cost itself turns into profit, such a product already has a little built-in margin.

However, what if failures occur epidemically after product release, with similar failure modes? It takes time to establish the needed corrective action because the exact cause must be researched and determined.

No matter the cause of the defect, it is not easy to find an adequate corrective, because there is not enough space on a densely designed structure to accommodate a remedy. It also takes a lot of time to prepare spare parts and the repair site. One rapid response to failure is to substitute an identical new product for the failed one, but the same problem will occur again if it is not fixed. If lots of products have been released into a widespread market before the issue is identified and resolved, it is a very risky decision to commit all the resources of the company to repairing the returned products, and difficult to recover the damaged company image resulting from a recall.

Moreover, all this work is usually done primarily by performance designers, who are rarely specialists in reliability. This should make a CEO wary. In fact, few CEOs understand the significant differences between the field of reliability technology and that of performance technology. CEOs imagine a bright future in which their newly developed product will prevail throughout the market once it is released, but there may be an ambush waiting for it in the form of reliability failure. Distressingly, many promising venture companies, as well as large firms, have disappeared due to reliability problems 2 or 3 years after a promising product's release. Many companies have had to shut down a strategic business unit that had exhausted its surplus and laid off its personnel due to reliability accidents. Because reliability concepts have mainly been explained with complicated mathematics, the ideas have been difficult for CEOs to grasp and apply.

Product failure is a physical problem. Let's focus on where the trouble occurs, or the failure site. Because every product is an aggregate of many structures, the unit structure of the failure site may only be visible when the product is disassembled. The loads applied to this unit structure produce stresses. Failure occurs when the stress is greater than the material strength, or when the material cannot endure the applied stress. Understanding this process is called *failure mechanics*, and its two elements are stress and materials. A desirable or reliable structure includes both well-dispersed stresses and reliable materials. Figure A.1 illustrates failure mechanics and its two elements.

This description sounds like it applies only to mechanical structures, but it is equally true of electronic devices, which have analogous structures. For example, semiconductors experience thin-gate rupture due to high voltages (electrical overstress (EOS)), and circuit wires have narrow-width openings due to high currents (electromigration). In either case, failure occurs when the materials comprising the structures do not survive the environmental and operational stresses applied to them; after all, every object from a microdevice to a high building has some architecture, and it is only as strong as its component parts.

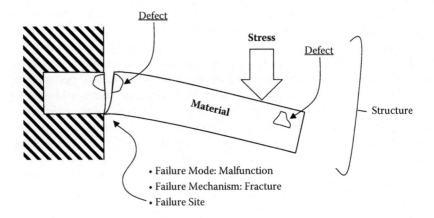

Figure A.1 Failure mechanics.

Because reliability is the relationship between stresses and materials, the solution for avoiding failure is altering the structures to better disperse stresses or replacing materials. But strong materials are not always reliable. Bronze is weaker than steel in mechanical strength, but better than steel in corrosive environments. It is best to select a material with consistently high mechanical strength throughout the product's usage lifetime.

In Figure A.1, the defect in the middle of the material will contribute to failure. Such defects are usually produced in the manufacturing process, which leads us to regard defects solely as a manufacturing issue. But the other defect, near the end in Figure A.1, will not lead to failure, which implies that the specifications for flaws should be revised and expanded to include such data as location and size.

Altering the structure or materials or making specifications more detailed assumes that achieving reliability pertains to design, not manufacturing. Since product design includes reliability, it is properly done jointly by the product designer and a reliability specialist. Consequently, design specifications should be classified into two categories—performance specifications and reliability-related specifications. When an item is inspected for a certain specification and turns out to be out of tolerance, the item will not work well if the specification relates to performance; if it is a reliability specification, the item cannot reach its designed target life.

In order to make products highly reliable, three kinds of specialists must work together organically: the product developer, the reliability engineer, and the failure analyst. The task of the product developer, sometimes called the "know how" specialist, is to improve a product's performance and decrease its cost. The reliability engineer is the failure mechanism specialist who identifies reliability issues lying dormant in the product. The job of the failure analyst, or the "know why" specialist, is to analyze in depth the failures identified, both inside and outside the company, and to clarify the real causes. The fields of these specialists are totally different. For a television set, for instance, the product developer is mainly an electronic engineer, the reliability engineer is usually a mechanical engineer with an understanding of failure mechanisms and parametric ALT (accelerated life test), which is the methodology for verifying whether the item's reliability meets those quantitative targets, and the failure analyst is a materials engineer.

Let's consider the ways in which they work together. Product developers, who understand competition in the market, pass information on the function and structure of the product, its reliability targets, and test samples to the reliability engineers. The reliability engineers complete tests and pass on the reliability indices (BX life, or lifetime reaching the cumulative failure rate of X percent, and annual failure rate within lifetime) and estimated expenses for future reliability problems to the product developers. They, in turn, pass on data about failure modes and failed samples identified in tests to the failure analysts. Failure analysts, using these data and the design materials from the product developers, perform their analysis, report it to the reliability engineers, and propose alternative design changes to the product developer.

The CEO's task is to functionally differentiate the organization accordingly. First, the product approval section should be divided into two parts—performance approval and reliability approval. The section for failure analysis should be set up separately.

The performance approval area manages both the usual and specialized performance tests, but hands over the tests that lead to material strength changes to the reliability approval section. Reliability engineers identify problems that might occur in the future, executing reliability marginal tests and reliability quantitative tests. Thus, this section should be furnished with environmental testing equipment and the specialized test equipment required for item approval tests. The reliability section should be able to configure specialized test equipment according to test specifications, as well as to design parametric ALT.

The failure analysis section needs both frequently used nondestructive and destructive analysis equipment and tools for measuring material characteristics. Since it is impossible to purchase all the necessary equipment, failure analysts should identify nearby universities or research institutes that have the kinds of high-priced equipment they may occasionally need and establish cooperative arrangements for utilizing their facilities. The functions of such facilities are diversified due to advances in computer technology and measuring speeds. Using state-of-the-art equipment, they can nondestructively visualize material defects in failed test samples, and with a focused ion beam, simultaneously remove a thin layer of material for further analysis.

Failure analysts should be sufficiently knowledgeable to consult with specialists outside the company in order to increase the accuracy of failure analysis. They need to understand all the processes of failure analysis in order to work with outside researchers in reviewing the characteristics, limits, and operation of various types of analysis equipment. Sometimes, manufacturers are asked to analyze the failure of problematic items, which is like setting the wolf to guard the sheep. Since the item manufacturer has a high level of technical knowledge about the item, its failure reports are generally accepted, but it would be better to understand these as simply useful information, because the manufacturer is not going to report results that might impact its company negatively. Thus, it is important for CEOs to establish their own failure analysis sections and control them internally, independent of any outside entity. To be trustworthy, failure analysis must be handled independently.

Mistakes in reliability engineering can create serious problems with new products and lead to ensuing claims from customers, with corresponding increases in service expenditures after product release. If failures are not analyzed in depth and correctly, chronic problems could remain in the manufacturing processes, and production expenditures will increase as inspections and other temporary fixes are implemented.

CEOs reconsider the essence of reliability design and furnish the organization to handle reliability-related issues, which has long been ignored.

A.2   Design technology is only completed with the establishment of approval test specifications

You don't have to be an engineer to understand the key technology systems involved in making a product. The hierarchy of hardware technology can be outlined, from bottom to top, in the following order: service technology, manufacturing technology, design technology, and origin technology. Service technology can be acquired from the service manuals available when purchasing products. Outlines of manufacturing technology can be found through examining service manuals and performing service activities. Detailed manufacturing technology can be ascertained by observing the activities of the original equipment manufacturer (OEM). But design and origin technologies are generally kept secret. While some of them are released to the public, the types of publications describing them, such as technical papers and newly publicized standards, are inadequate.

Combining service technology as part of manufacturing technology and considering origin technology as a prior stage to design technology, hardware technology really only comprises two types: design technology and manufacturing technology. The former is used to design the

Table A.2 Design Technology and Manufacturing Technology

Description	Design technology	Manufacturing technology
Concept	Product design Developing specifications for performances and manufacturing	Production according to specifications
Type	Basic theory Technical papers Testing and analysis reports Market research data	Process sequencing technology Process configuration technology Quality system
Difficulty	Difficult to understand	Easy to follow
Field	All areas of science and engineering	Mainly mechanical and electrical engineering

product and decide the specifications for its performance and manufacture; the latter creates products matching those specifications. Ultimately, the hardware product is the result of manufacturing according to design specifications.

Manufacturing technology encompasses process sequencing technology, the process configuration technology requisite to each sequenced process, and the quality system; these are mainly developed and implemented by mechanical engineers. The outcomes of this technology can be seen in the production factory. But design technology is much more complex. Deciding on specifications requires working with high-level knowledge, because the critical information for creating a good design resides in a variety of fields and levels of knowledge. This information is subsequently reported via many instruments: basic theory studies, technical papers, testing and analysis reports, market research, and so on. Integrating these widely diversified materials has to be done using a multidisciplinary approach. Moreover, to secure a competitive advantage, companies must develop unique proprietary techniques from basic scientific research, rather than relying solely on commonly used technologies. Consequently, it is not easy to understand design technology. Table A.2 summarizes the distinctions between design and manufacturing technologies.

Where does the complicated design technology come from? It originates in the minds of specialists, then is made available in technical articles, specifications, reports, and drawings; finally, it is realized in actual, properly produced items. The relevant technology can be obtained by hiring experienced experts, by studying related specifications and documents, and by measuring and analyzing manufacturing equipment with high precision. Being able to produce world-best products implies that a company has acquired excellent technology in these three modes.

Table A.3 Results of Design Technology

Order	Description	Results
1	Design materials	Bills of materials, drawings, specifications of constituent materials, components, and units
—	Production equipment	Manufacturing facilities
2	Approval specifications	Measurement specifications of performance and performance fundamentals; testing and inspection specifications of assembled products, units, and components
3	Verifying equipment	Facilities to measure and inspect performance; equipment to test and analyze quality

Design technology can be clearly understood by reference to this model. A design results from the expression of ideas conceived by scientists and engineers, published in papers, and visible in actual products, in main activities of production and verification. This process results in two essential systems, subdivided into four categories: production papers, production equipment, verifying papers, and verifying equipment. Production papers include design materials, bills of materials, drawings, and the specifications of constituent materials, components, and units—the first results of the design process. Production equipment includes the manufacturing facilities that produce parts and assembled units, which are mainly purchased from outside suppliers. (This pertains to manufacturing technology and thus is excluded from the results of design technology.) The second set of results of the design process includes verifying papers that provide approval specifications, such as testing and inspection specifications for completed products, units, components, materials, and so on, and measurement specifications for performance fundamentals. These include testing and measuring methods, sample size, sample processes, and acceptance levels, and therefore should be based on statistics. Finally, since these approval specifications must be actualized in hardware facilities, configuration equipment technology must produce verifying equipment that measures performance, checks qualities of relevant items, and so forth—the third set of results. As shown in Table A.3, the results of design technology are thus design materials, approval specifications, and verifying equipment.

Holding the design technology for a certain item means that someone in the company establishes design materials and approval specifications, and contrives verifying equipment. If it is impossible to measure or verify the performance and quality of products precisely after they are made, not all the relevant technologies have been addressed. Many countries designing televisions, for example, use various instruments made by the

United States, Japan, or Germany, which means the companies outside those three countries are not developing their own original technologies. If, say, there is no original apparatus for checking picture quality quantitatively as one of the performance fundamentals, the company does not have the differentiation technology to make its products unique. This will damage the company's ability to survive in worldwide competition.

The final stage of technology is measuring and testing. It is very important to confirm quantitatively the value of a product's differentiation from competing products. Establishing verifying specifications and devising equipment that measures, both precisely and quickly, performance and product lifetime requires acquiring more advanced technology. It is desirable and necessary, although the challenges are great, for a company to develop its own such equipment because available instruments that are merely similar cannot accurately assess its proprietary technology. Finally, thorough and accurate evaluation specifications and facilities are as important to the final product as original design.

Designed products should be perfected by measuring characteristics and testing lifetime, conforming the differences to the targets, and complementing them with various measuring and testing equipment according to related specifications. Therefore, world-class products designed by a certain group must be completed by their own in-house approval section, establishing related specifications and designing facilities according to their specifications.

Now, let's consider the concept of design approval specifications. Product performance should match the designer's intentions and continue consistently throughout the product lifetime expected by the customer. The intended performance is confirmed by its adherence to performance specifications, and checking its invariability during use is done through reliability specifications. As new products are introduced with different performance expectations, the specifications must be revised; if the structure and materials differ from those of previous products, the reliability specifications must also be altered. Because new products for the prevailing market naturally incorporate many changes, the "pass" obtained from the specifications for previous products will no longer be acceptable; obviously, the specifications applied to the old product may be inappropriate for new ones. Since the starting point of the process is valid specifications, the specifications for new or altered products must be reviewed and revised. This is just the second set of results of design technology.

In the 1980s General Electric (United States) and Matsushita (Japan) switched from reciprocating to rotary compressors for household refrigerators. Although a major alteration had been made in the basic structure of the refrigerator, they did not review all the related specifications. Instead, they simply accepted the previous good results obtained in accordance with the specifications for reciprocating compressors and launched a new refrigerator model with a rotary compressor into the market. A few years later, due to the malfunction of rotary compressors, many refrigerators were replaced with the old type, resulting in enormous financial losses. The manufacturers had assumed that the refrigerator remained essentially the same and could be checked through existing specifications—even though a core item was different. Generally, an altered product requires at least a different fixture to attach it to a tester, but in this case no changes were made in the testing. This was a fairly standard practice.

Company G has established a life test specification that runs continuously for 2 months and supposedly simulates 5 years' actual use. With this specification, 600 compressors were tested and not a single failure was detected. Consequently, over 1 million new compressors were released by the end of 1986. Within a year an avalanche of failure was reported. Although the materials had been changed from cast iron and steel to powdered metal in a completely different structure, reliability engineers thought the existing test specifications would still be appropriate to test compressors incorporating these new, cost-saving materials. The great number of test samples, perhaps, made them comfortable about reliability, but the resulting failures within a year indicate that the existing test specifications, anticipated to prove at least 5 years of operating life, were completely inapplicable for the new compressors. This highlights the importance of confirming the validity of reliability test specifications.

Company M's refrigerators experienced epidemic failures over several years after the release of the new rotary compressors that incorporated design changes for cost reductions. Reliability engineers should have realized that minor design changes and even trivial deviations in the manufacturing process could greatly influence the reliability of rotary compressors, which have a relatively weaker structure than reciprocating ones. They could have identified the problems in advance if life tests were conducted on whether the B1 life exceeded 10 years and if failure rate tests were performed with an adequate sample size for the target. This points to the necessity of careful configuration of reliability test specifications, both to meet the target and to confirm the validity of existing specifications.

There are significant differences between the structures of rotary compressors and those of reciprocating compressors, though they produce equivalent functions. Let's enlarge our understanding of the structures that change in accordance with the two elements of failure mechanics, stress and material. The rotary-type compressor has a longer gap between sliding metal components of various materials, along which high-pressure gas leaks can occur, than does the reciprocating type. This gap makes it difficult to produce high compression. This is why rotary compressors are used for air conditioning systems, which require only cool air, and reciprocating compressors, with their shorter leakage lines and greater compression, are incorporated into refrigerators, which need freezing air. Of course, rotary compressors can produce very cold air, but the technology to do that is highly exacting, which means that there are more critical to quality (CTQs) than for a reciprocating compressor. Note that, in this case, it is crucial to find and manage all the attributes influencing leakage, hampering lubrication, and causing wear due to the higher compression ratio and the longer leakage line.

In over 30 years of manufacturing experience, I have never heard anyone express the opinion that there should be a specifications change in this case. But let's think about that. It is simply common sense that once the product is changed, the specifications should be revised. How can we apply the same specifications to a refrigerator with a new part? We cannot. Thus, all the specifications should be reviewed when a product changes, and changed to the same extent that the hardware has changed. The more change, the more revision.

Our tasks should be to review and complement reliability specifications adequate to new products, and sometimes to establish new specifications and to configure equipment. CEOs should put their efforts into the completion of approval specifications through a newly motivated reliability organization.

A.3   The core concept of the approval test is the quantitative estimation of the product

If the material undergoes large stresses, it will break; if it receives small stresses repetitively, it will also break in the end. Breakage due to large stresses can be detected without difficulty, but failure due to the accumulation of small stresses cannot be easily detected in a moment, so lifetime tests are required confirming whether the time to failure is within the lifetime. As noted, business survival would be impossible if products demonstrated epidemic failures in 2 or 3 years under normal environmental and operational conditions—especially for durables, which are expected to provide over 10 years of failure-free use. Thus, every unit within a finished product should be tested in order to estimate its lifetime quantitatively in terms of possible failure modes.

Quantification should be the basic datum of all judgments. The pros and cons about new product development can be debated endlessly between the design section and the assurance section with nonquantitative data. Assurance personnel want to ensure that the product will not fail due to failure mode revealed in the in-house test. Design people reply that corrective action is unnecessary for a failure that occurs under severe, abnormal conditions, because such conditions are not likely in real use. It is also difficult for CEOs to blame the designers for such failures, given that they struggle to meet scheduled, often short, timeframes under pressure to meet the assigned cost limit. If CEOs were aware of these designer assurance section arguments, they would be forced to make policy decisions to resolve the issue. As discord between the two sections increases, CEOs sometimes order the section chiefs to change positions with each other to promote mutual understanding or to commit the task of product approval to the product development chief. Or perhaps the CEO moves product approval to a higher authority than the section chief, hoping for staff harmony and a trouble-free product. But when one group is favored by the CEO's policy decision, the morale of the people recommending the other approach is compromised. These are all band-aid solutions, and real problems cannot be solved with them. The product is put at risk when organizational harmony is the prime decision factor. Such decisions are rational, but most of them are not made on scientific bases. In order to avoid these problems, the CEO should try to make such decisions on objective bases. These issues can only be solved by using quantitative data developed by building reliability technology capability.

The task of establishing new quantitative test specifications adequate for innovative items seems quite daunting. But not all the old specifications need to be changed. After classifying and synthesizing all the test specifications available as of now, let's reconsider and quantify them. Most new products are related to current models, which often have lingering intractable problems. First and foremost, whether these extant issues occur in the new product should be determined. Next, the potential issues due to the newly designed portion should be predicted. Finally, issues of comparative disadvantage with competitors' products should be reviewed. These three kinds of issues should be checked one by one and divided into two groups: performance issues and reliability issues. If an issue is related to materials rupture or degradation, especially over time, it is a reliability issue; if not, it is a performance issue. The specifications relevant to any problems must be explored and revised.

Performance issues should be further divided into common performance issues and special (environmental) performance issues; doing this enables approval engineers to consider all necessary tests. Special performance issues occur under unusual environments, such as elevated electromagnetic fields. The necessity of a certain performance test can be checked easily if the environmental conditions under which the items function are investigated in detail.

Reliability issues can also be divided into two groups: reliability marginal issues (unusual environments) and reliability quantitative issues (lifetime and annual failure rate within lifetime). Reliability marginal issues include physical changes in items under severe or peculiar conditions, including unusual usage environments, such as electrostatic overstress, lightning surge, and storage and transportation. Reliability quantitative issues include reviewing and estimating lifetime and annual failure rate within lifetime under almost normal conditions (this will be the measure to the problem of company M, discussed in Section A.2). Most existing test specifications can be classified as reliability marginal tests, and reliability quantitative tests can be set up as new tasks in a test plan of BX lifetime. Figure A.2 illustrates this classification.

Now let's consider actual tests for new product approval. As usual with tests that check performance under normal conditions, the sample size can be small unless it is necessary to confirm the average value and the standard deviation of performance characteristics distribution. With special performance tests and reliability marginal tests, it is enough to sample a few items or test for a relatively short period. When the probability of peculiar environments is low and the affected numbers of products in such instances is low (with the cost of reimbursements for failed products correspondingly low), the typical pass/fail test is sufficient. But in the case of reliability quantitative issues, setting up the test specifications is quite different. Careful preparation is necessary in order to calculate an item's lifetime. Testing many samples or doing so for a long time is not the same as a lifetime test. Quantitative information results from scientific test design, such as the establishment of severe conditions responding to

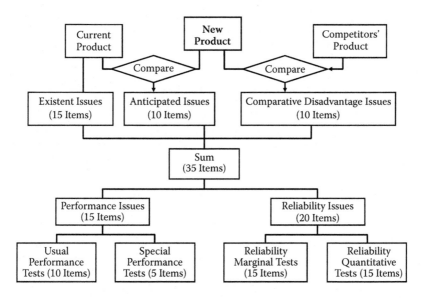

Figure A.2 Conceptual framework for quality assurance.

actual stresses, computing the test duration to assess material degradation, and calculating a sample size adequate to determine the BX lifetime target. Furthermore, testing should be designed to produce failure with the same failure mechanism that affects the actual product, reviewing in detail the parameters related to the applicable stresses. The results should be expressed as the BX life Y years, which means that the period to reach the cumulative failure rate of X percent is Y years.

Let's consider the outline of quantitative test specifications, using common sense and mental arithmetic about the case of rotary compressors in Section A.2. (You can find its basis in Chapter 5 of the main book.) If the lifetime target is B1 life 10 years, it is satisfactory if there are no failures in a test of 100 compressors (that is, the cumulative failure rate is 1%) tested for 4 months (120 months/30) under severe conditions leveled up by an acceleration factor of 30. Note that the sample size can be decreased to approximately 25 if the test period is extended to two times the period, or 8 months. For a failure rate target of 0.3% for 3 years for all CTQs, 330 compressors would be tested for 2 months (36 months/20) under severe conditions, leveled up by an acceleration factor of 20, with the condition that no failures are found. If there is concern about overstress failure due to material degradation, the above test would be changed, for a failure rate target of 0.5% per 5 years, to 200 compressors tested for 3 months (60 months/20) under severe conditions, leveled up by an acceleration factor of 20, with the condition of no failures being found. In the case of failure rate confirmation, secure random sampling would help to find various problems corresponding to material defects and process deficiencies. Now let's figure out the test conditions. It approximates actual usage to assume that compressors are operating intermittently, which causes them to work in a state of insufficient lubrication, with increased pressure differences under higher temperatures.

If CEOs grasped this concept and could mentally interpret the test reports, they could understand the framework of test specifications and assess their validity. If they assume their engineers find and fix reliability failures, they feel comfortable that the test specifications are effective; if there is no failure found, they can request complements to the test specifications, such as retesting, extending the test period, or leveling up the severity of the test conditions.

When I worked for company S, the product quality in a certain division had not been particularly improved for over thirty years in spite of various attempts to advance it. After a parametric ALT design specialist was developed, the product quality rapidly improved through his reliability validation. This was enhanced greatly by a company-wide change in the product development process. Validating by parametric ALT and achieving a quantitative confirmation of reliability, in addition to conducting existing pass/fail reliability marginal tests, had newly been incorporated into the product development process. Although, of course, there were other excellent changes in aesthetic design and constant efforts to improve product performance, the success of the quality improvement program could be ascribed to developing designs using newly established test specifications for product lifetime. With them, the annual failure rate decreased and product lifetime increased. At that time, I executed a BX life test on every item that dramatically improved the failure rate of core items, such as compressors and ice maker units. After 1 year, the outcome was already apparent. In a study of major home appliances in 2005 by J.D. Power and Associates, company S's refrigerator scored 817 points on a customer satisfaction index, based on a 1,000-point scale, ranking first—22 points more than the next competitor, and 45 points more than the third-ranking company. Moreover, the side-by-side refrigerator received greater public favor than the highest-ranking dishwashers and ovens, scoring 50 points more in customer satisfaction.

In order to beat competitors in a globalized market, a corporation should produce products at a similar or lower price, with higher customer satisfaction in important attributes, and superior advantage in core performance. The more design changes are needed to do this, the higher probability there will be of unanticipated problems. Until now, CEOs are always worrying about failure problems. But we can assure that these problems will disappear using newly established specifications, including parametric ALT, together with scientific failure analysis. This frees us to apply fresh ideas to our products. Further, in order to be first in the marketplace, we can concentrate on researching and developing original technology to elicit customers' favor.

CEOs should accept reliability quantitative approval as the core activity of product approval in the process of developing a product for great success.

Index